The Inheritance of Coat Color in Dogs

The INHERITANCE OF COAT COLOR *in* DOGS

By CLARENCE C. LITTLE

Director Emeritus of the Jackson Memorial Laboratory

Bar Harbor, Maine

ς

1979—Seventh Printing

HOWELL BOOK HOUSE
230 Park Avenue
New York, N.Y. 10017

First published 1957

Exclusive Publisher and Distributor in the United States and the World—Howell Book House Inc.

PRINTED IN THE UNITED STATES OF AMERICA
ISBN 0–87605–621–4
LC 67–8658

Preface

THIS book has been made possible by a number of people and organizations. Acknowledgment is gratefully made to the Trustees of the Rockefeller Foundation for their basic support of the Jackson Laboratory program of genetics in relation to behavior. This support provided the opportunity to breed the animals used as primary data and to maintain the services necessary to the recording and interpretation of the facts obtained. Especially do I wish to mention the vision and support of Dr. Alan Gregg as director of the Division of Medical Studies and of Dr. Robert Morison, his successor in that field of the Rockefeller Foundation's program.

Thanks are also due to the directors of the American Kennel Club for the aid of that organization in distributing cooperators' blanks and to the thousands of breeders throughout the United States who submitted data on those blanks.

During the inception and preparation of the manuscript my associate Miss Edna M. DuBuis, senior research assistant at the Jackson Laboratory, followed each step in detail. Her great store of knowledge on sources of information concerning dogs and her patience in seeking, revealing, and discussing new sources have been invaluable.

I am also appreciative of the cooperation of my other associates, especially Doctors J. P. Scott and J. L. Fuller, whose records of experimental animals have contributed a part of the data here given.

In the technical preparation of the manuscript acknowledgment is due to Miss Geraldine Canning and Mrs. Marjorie Higgins for stenography and proofreading and to Messrs. Philip Montella and George McKay, Jr., for the photographic work in connection with the illustrations.

CLARENCE C. LITTLE

Contents

Contents

Contents

Contents

Contents

PART ONE

*Introduction and
Genetic Background*

CHAPTER I ✎

Introduction

THIS book discusses our present knowledge of the genetics of coat color in dogs. It recognizes both the long duration of man's contact with dogs and the degree to which early records of the color and conformation of domesticated dogs are lacking in completeness. A considerable amount of new data derived from sources described in the section entitled "Materials and Methods" is presented and is considered in the light of earlier analyses.

It is highly improbable that any one breeder or institution will provide numerical data on dogs based on direct observation equal either in quantity or in quality of detail to those available for small laboratory rodents. An attempt has been made, therefore, to utilize data from other sources, with due emphasis on the contrast between these data and those derived from direct observation.

It is hoped that a careful study of the results obtained will enable breeders who are interested in the scientific aspects of the subject to derive helpful and useful information, not only on the production of the color types desirable from a fancier's point of view, but also on the avoidance, elimination, or proper scientific utilization of unusual color types or of those not desired by breeders.

With this in mind the present book consists of three main sections. Part One consists of a discussion of material and methods from and by which the data were derived. This is followed by presentation of general genetic background material to cover the principles of Mendelian heredity. In Part Two a more or less technical discussion of the genetic data has been presented under the various gene locations involved. Books and articles cited in this part have been chosen from types of study similar to the present one.

Part Three is a brief analysis by breeds, with less technical emphasis and with references to sources of information of a type likely to be known by the practical breeder. Even with a desire to be of service to the breeder, however, it has not seemed wise to use nontechnical language entirely or to avoid reference to the gene symbols used earlier. Breeders who desire to utilize existing scientific knowledge of genetics will need to spend some time in serious study of the symbols and methods employed in that science. If this familiarity with genetic tools were not necessary, many problems of coat-color inheritance would have been solved long ago.

Not all breeds recognized in this country have been covered in this text, but an attempt has been made to include the more popular and important breeds. The breeds omitted are those for which data on coat-color inheritance are considered to be inadequate.

Liberal use has been made of a number of very helpful and valuable texts. Some of these, like Dawson's excellent summary review (1937), Whitney's general discussion of dog genetics (1947), and Burns's very useful recent book (1952), are recommended to the reader. Perusal of them will give a fair picture of how much of the present volume is confirmatory and how much represents new interpretation. Winge (1950) made a serious and praiseworthy attempt to present a discussion of coat-color genetics in dogs, but his book is not so clear and useful for general reference as the other texts just mentioned.

A number of special articles which should be helpful to certain breeders will be mentioned under the sections on particular breeds or under sections on specific loci.

It is hoped that the new data and the comparative analyses in the present book will prove useful to geneticists and dog breeders. Above all else the author desires that the present book may serve as an impetus to further research regardless of whether or not there is agreement with the views expressed herein.

MATERIALS AND METHODS

Data included in the present volume are derived from three sources: direct observation, breeders' records, and the literature on the incidence and inheritance of coat color. Some discussion of each of these types of data will be helpful.

Direct Observation. Dogs have been studied at the Jackson Laboratory without interruption since 1941. A colony of between 200 to 225 animals of a number of breeds is constantly under investigation. Over 4,100 puppies have been produced and recorded. They are the result of matings planned for specific purposes in relation to the genetics of color and other morphological traits or in relation to certain behavior characteristics.

The matings have been within the breed in some cases and between breeds in others. The former serve the purpose of maintaining standard types for use in genetic analysis. The latter provide a test of similarities and differences existing in animals of the same color appearance (phenotypic resemblance) in the various breeds and their hybrids. Only by studies based on the two types of matings can the extent to which any explanatory hypothesis applies be determined.

All animals at the Laboratory are produced and reared under essentially similar conditions in respect to housing, diet, and other care. This is an attempt to reduce to a minimum the variation which may be caused by environmental factors.

The breeds used in these studies at the Laboratory include the following: Afghan Hounds, Beagles, Dachshunds, Greyhounds, Norwegian Elkhounds, Great Danes, Doberman Pinschers, Shetland Sheep Dogs, Fox Terriers, Irish Terriers, Kerry Blue Terriers, Scottish Terriers, Pointers, English Setters, Irish Setters, Cocker Spaniels, English Springer Spaniels, Chihuahuas,

Schipperkes, Chow Chows, Dalmatians, Poodles, Pugs, Basenjis, Boxers, Great Pyrenees, Weimaraners, and Collies, a total of twenty-eight.

Personal Experience. The author has owned and bred dogs since childhood. For over twenty-five years he had contact with the kennels of his father, James L. Little. This kennel, which the author took over in 1914 upon his father's death, consisted of a varying population of from 20 to 100 dogs, chiefly Clumber and Cocker Spaniels, Brussels Griffons, Dachshunds, and Scottish Terriers.

As a judge at dog shows, the author has had experience with all Hounds, all Terriers, Setters, Pointers, all Spaniels (except the toy varieties), and Old English Sheep Dogs.

Breeders' Records. The breeders' records used are of two general types: (1) those collected on specially prepared cooperators' record blanks issued by the Jackson Laboratory, and (2) those obtained in correspondence with individual owners or with kennels.

Cooperators' Blanks. These blanks, the two sides of which are shown in Figs. 1 and 2, were prepared at the Laboratory and distributed by the American Kennel Club to those registering purebred dogs. Additional blanks were forwarded from the Laboratory at the request of individuals who used the first one received.

It will be noted that the blank, when properly completed, gives information on many important facts concerning the complete litter recorded.

In tabulating and utilizing these data, we have discarded imperfect or carelessly recorded blanks. It is not difficult to make a selection between careful and accurate correspondents and others.

There are, however, certain difficulties, unavoidable under the present methods of describing and registering dogs, and these should be mentioned. One of the most persistent is the confusing and nonscientific terms used in recording color in cer-

Introduction

Roscoe B. Jackson Memorial Laboratory Cooperators Blank No._____
(LEAVE BLANK)

Please fill out and mail to the Roscoe B. Jackson Memorial Laboratory, Box 847, Bar Harbor, Maine

A. K. C. No. DAM _____ A. K. C. No. SIRE _____

BREED _____ BREED _____

COLOR _____ COLOR _____

DATE OF BIRTH _____ DATE OF BIRTH _____
of Dam   Month   Day   Year   of Sire   Month   Day   Year

DATE OF BIRTH OF THIS LITTER _____ IS THIS 1st, 2nd, 3rd, etc. LITTER OF BITCH? _____
Mo.   Day   Yr.

PUPS BORN ALIVE — Use One Line For Each Pup				PUPS BORN DEAD — Use One Line For Each Pup			
No.	Sex	Color	Note Any Peculiarities of Structure or Color	No.	Sex	Color	Note Any Peculiarities of Structure or Color
1				a			
2				b			
3				c			
4				d			
5				e			
6				f			
7				g			
8				h			
9				i			
10				j			
11				k			
12				l			
13				m			
14				n			
15				o			

General Information

1. This record will help in important scientific research and will be greatly appreciated by the Roscoe B. Jackson Memorial Laboratory, Bar Harbor, Maine, which is working in cooperation with the American Kennel Club in research on heredity in dogs.
2. Accuracy is important. Incomplete or incorrect records give misleading results. Include all pups – normal and abnormal.
3. In the use of this information no breeder or individual dog will be identified so that recognition is possible.
4. Use one line for each pup. Where the number born alive exceeds the lines available use vacant lines under the heading "Pups Born Dead" but clearly note that pup was born alive. Use extra paper when longer descriptions are desired.
5. Pecularities of form or color (including unusual white spots) are of special interest. Use outlines on back of this sheet. Identify each outline used with number or letter corresponding to front of sheet. Presence of extra toes, dew claws, kinky or short tail, "overshot" and "undershot" jaws, eye and ear defects should be noted.
6. Peculiarities of behavior or temperament which do not seem to be the result of training or injury should be noted. Examples: – Nervous movements, unusual aggressiveness or shyness, lack of maternal care, etc.

Please Fill In Lines Below
(Please Print)

NAME OF BREEDER _____

ADDRESS _____

Remember That Correspondence Or Questions Are Always Welcome
Questions of a Veterinary Nature should be referred to a Veterinarian.

Fig. 1. Cooperators' blank (front).

7

DIRECTIONS—Place the number or letter of the pup (corresponding to that used on the opposite side of this sheet) in the upper left hand corner of a square in the blank after "No.". Copy the color pattern of the pup, first for the right side, then for the left. Be sure to indicate colors clearly. Anything which cannot be shown on the drawings may be written under "notes" below.

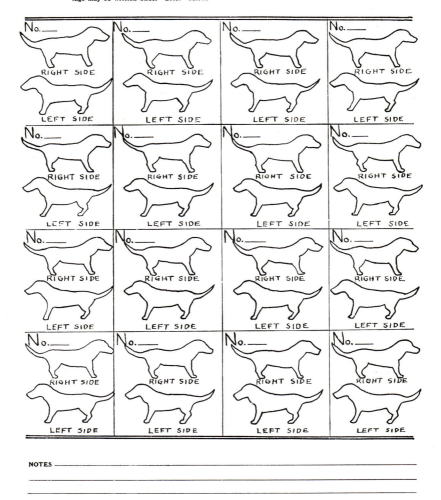

NOTES _____

Fig. 2. Cooperators' blank (back).

tain breeds. DuBuis (1948) has commented on this in a brief note. Wherever such a situation is found, no attempt has been made to differentiate between grades or shades of a given color. Thus in Cocker Spaniels, where the maximum of personal preference and the minimum of consistency in color description have been encountered, all shades of the red-yellow series have been lumped together in analysis and use of the data.

Another unavoidable source of inaccuracy is lack of certainty that, when a bitch is shipped for service, the dog reported to have been used was *actually* used. Inaccuracy of this sort may result from intent or from accident. Indeed, the latter may occur whenever isolation and observation of the bitch are not complete and certain.

There may also be errors due to carelessness in recording a large litter of puppies of more than one color. If, however, blanks are selected for use on the basis of neatness, accuracy, and completeness, this source of error should not be of major importance.

In some breeds coat color may change from birth to maturity or even throughout the whole life span of the individual. In the development of coat colors which have a durational aspect, it is evident that observations and records made at any one period during the lifetime can be at best only a part of the whole story of the action of the gene or genes involved.

An additional factor to consider is the difficulty of detecting, even with a thorough examination, white spots on a pale cream dog or the converse—cream-colored spots on an otherwise white dog. If one is dealing with a breed whose standard calls for "white," it is very human not to hunt for, emphasize, or even mention the presence of minute and possibly inconspicuous spots of pale color.

Even with all these handicaps, data from breeders have proved valuable in the analysis of color genetics. When the number of individuals is extensive and the categories used in analysis are sufficiently broad, as is the case with most of the breeds reported here, a great deal of suggestive and confirmatory information will be obtained.

PERSONAL CORRESPONDENCE. Many breeders, on receipt of their cooperators' blanks, have written the Laboratory describing some interesting and unusual variation which has occurred in their kennel. Often these breeders have sought an explanation of the observed variation or have offered to carry out matings planned to learn more about its origin or genetic behavior. Contact has been maintained with such persons, and, wherever it seemed justifiable, their experience has been utilized.

Information has come from a number of individuals who, for varying lengths of time, have been actively interested in their own observation and analysis of the method of inheritance of coat color in various breeds. Their background of training in genetics varies considerably, as does the degree to which their point of view is affected by their desire to prove or to disprove certain hypotheses. It is therefore necessary to evaluate their contributions according to their degree of impartiality or lack of prejudice.

LITERATURE

Here again one encounters material which differs greatly in its origin, extent, and reliability. Obviously, publications on the genetics of coat color in dogs utilize data obtained from the various sources listed above. Many short communications in dog journals deal with exceptional color variations. These, of course, have the same weaknesses that personal communications directly to the Laboratory possess. Yet they give interesting evidence of the appearance of possible mutants and of the extension of the range of variability in coat color in different breeds.

Of less value are the longer and more discursive articles in nonscientific journals attempting to apply such biological or mathematical principles as Mendel's law or the effects of inbreeding. With very few exceptions, the authors of these articles have a regrettable lack of balance between experience as breeders of dogs, on the one hand, and experience as theoretical geneticists, on the other. In the frequent confusion regarding the application of Mendel's law and the limited knowledge of the law's scope and modifications apparent in these communications, there is discouraging support of the popular saying, "A little learning is a dangerous thing."

Scientific journals print many short papers based on direct observation of a distinctly limited kind or on more extensive data derived from stud books or similar sources. For the most part these represent conscientious efforts to compensate for the limitations due to lack of time and funds which must always accompany the method of *direct* observation on any large-scale basis. The data contained in these short communications are of value, and many of them have been considered and included either directly or as background material in the present publication.

In addition to the above, there are the longer and more comprehensive texts mentioned in the Introduction. The author has been faced with the necessity of selecting the best method of considering and incorporating the data and conclusions presented by these works. One method would be to offer detailed discussion, comparison, and criticism of their data in the light of those presented for the first time in the present volume. This would entail a great many specialized terms and a technique which almost certainly would produce "hard reading," constant need for direct comparison, and probably obscurity and confusion, in spite of the best intentions and most careful efforts. Another method would be to present the hypotheses of coat-color inheritance advanced by each author, together with the data for their support, and to leave broad comparisons with the interpretations of other authors to some future time and occasion. This would confine our references to these major contributions to those points where specific matters can be profitably discussed and compared. It has seemed to the author that the latter type of approach, combined with some consideration of analogies and possible homologies between the coat-color types and patterns of dogs and those of other laboratory mammals, would be the better of the two alternatives.

On the subject of dogs in comparison with other laboratory mammals, little has been added to the literature since the excellent short paper by Haldane (1927). Since then new mutations and new data have appeared, some of which will be included in the present study.

CHAPTER II ✑

General Genetic Background

EVERY dog owes his coat color to the combined activity of a number of units of heredity, or genes. Some of these have recognizable major effects; others are as yet unrecognized as individual influences. Those at present identified will be briefly described, and each of them will then be considered in some detail.

The gene, according to the overwhelming majority of students and investigators of heredity in mammals, has a characteristic chemical composition and a configuration which reproduce themselves accurately and identically in each generation of germ cells. Exceptions to this rule are sufficiently rare as to constitute an event of scientific note and to deserve a special name—*mutation*. Once such a change, or mutation, occurs, however, the new form of gene perpetuates itself with the same jealous tenacity as did the original form.

The genes are located in linear order, like a string of unequally spaced beads, in microscopic structures known as chromosomes. Every normal cell of every individual of every species of mammal has a fixed and characteristic number of such chromosomes. Each mature sex cell ready to engage in the process of fertilization has one-half of the number of chromosomes found in any other type of cell in the same individual. The number in the mature sex cell is called the n number; that in all other cells, $2n$. In dogs the n is 39 and the $2n$, 78.

Chromosomes can be seen in cells prepared for study under the high power of compound microscopes. They can be identified individually under the compound microscope and the phase microscope or by other instruments that give high magnification or clear definition of extremely minute particles.

When studied in this way, the chromosomes within each sex cell are found to differ in size and form consistently and characteristically, so that they can truly be called "individual." These differences in size and form continue faithfully and persistently through succeeding cell generations.

In all cells except the mature sex cell there are pairs of chromosomes. In each pair the two members are identical [1] so that each body cell is essentially a dual structure, with duplicate chromosomes.

In each mature sex cell only *one* member of each and every chromosome *pair* is found. The *number* of chromosomes has been reduced by one-half, but each individual chromosome type is still present. The biological and genetic reason for this reduction is clear. When the male and female sex cell, each carrying one-half the body cell, or *n*, number of chromosomes, combine at fertilization, the original double, or *2n*, number will be regained.

Although special techniques have made it possible to distinguish chromosomal structures corresponding with the known, or expected, locations of individual genes, there is no recorded case in mammals where any *visible* size or structural difference was discernible between mutant and parent gene types.

Each chromosome has its own content of genes which belong to it and remain with it. Between *different* pairs of chromosomes there should and usually does exist complete independence. A gene in chromosome *1* will, therefore, show no relation to, or association with, a gene in chromosome *2* in the method of or during the process of inheritance. On the other hand, if two different genes are located in the *same* chromosome, there should

---

[1] An exception to this rule is found in the pair of chromosomes particularly influential in determining the sex of an individual. In this case the two "pairing" chromosomes are unequal in size and shape.

and does exist a relationship, or linkage, between them in inheritance. The existence of such a linkage may be detected by constant departure from the ratio of types expected by the normal independent recombination of Mendel's law.[2] *The genes in the same chromosome tend to remain together.*

In the basic coat-color genes of dogs, which we shall discuss, no evidence of linkage has as yet been published. For the present, therefore, we may treat them as being independent, each gene base or *locus* (plural, *loci*) being in a different chromosome.

As data on color inheritance in dogs become more extensive and more frequently based upon deliberately planned and directly observed matings and progeny, we shall in all probability find evidence of linkage between some of the gene pairs now known. Until that time, however, the hypothesis of genetic independence is the proper working basis.

In the simplest form of heredity according to Mendel's law there occurs one type of gene that comprises both members of a gene pair. As long as this simple form continues, no experimental procedure will detect the existence of the gene pair, because the superficial appearance of *all* animals will be identical. It is only when one of the gene pair mutates that a visibly distinguishable new color variety is produced and by its behavior in inheritance enables us to identify the existence of its previously indistinguishable partner.

The various forms of a gene which can occupy the same site (locus) in a chromosome are called *alleles* of one another. *Allele* is a contraction of a term formerly employed, *allelomorph*, which means "another form of." This terminology therefore retains

---

[2] Clearly, in order to demonstrate the mathematical significance of such deviations from normal mendelian distribution, single litters or small numbers will not suffice. There are well-established mathematical procedures to measure what the probability is that the deviations from independent recombination are due to some cause other than chance alone. Usually a probability of 30 : 1 against "chance" is considered highly suggestive, and one of 100 : 1 or more as significant.

the concept of basic similarity, but of different arrangement, of the component parts—in other words, a new form of a mutant gene compared with its original form.

When more than two alleles have been identified at a single locus, they are called *multiple alleles*. In the dog four loci for coat color have been identified which may contain any one of four alleles, and six which contain two each.

Whenever three or more alleles are found, they can usually be classified and arranged in the order in which they can mask, conceal, or cover the expression of the other members of the series. The word *epistatic*, "placed above," is applied to a gene that conceals one or more others, while the term *hypostatic* denotes "standing under," and is applied to the concealed member of the series. Thus in a series of three there will be a top-ranking member, epistatic to both the others, a middle member, hypostatic to the top member but epistatic to the other member, and the bottom member, which is hypostatic to both its partners.

In the ordinary pair of Mendelian units the one of the pair which masks or conceals the other is called *dominant*, and the one which is masked, or concealed, *recessive*.

It is important to recognize and to remember that such masking or concealment (dominance) is often *imperfect* or *incomplete*. When this is the case, an animal possessing the *two* members of a gene pair shows visible and simultaneous development of characteristics dependent upon *each* of them. This may also be true of the epistatic-hypostatic relationships of any two members of a multiple-allele series, just as it is of the dominant-recessive relationship in a single gene pair. In each case either the recessive member of a single gene pair or the bottom member of a multiple-allele series should, when two animals characteristic of that type are mated, produce *only* animals similar to themselves with respect to the gene in question. This will become evident as the experimental data are presented.

The only exceptions will be when a mutation occurs. This,

as before stated, is a sufficiently rare event to be worthy of scientific note and record whenever it is encountered.

It is hoped that the reader will be able to absorb the principles of Mendel's law by study of the results of the matings recorded in the present work. If, however, direct contact with the theories underlying, and the data supporting, this law is desired, it can be obtained from *The Physical Basis of Heredity*, by T. H. Morgan (1919), chapters 1–5.

The research by which Mendel's law was discovered and amplified and by which it is applied represents careful work by brilliant and industrious research workers. The data on which the law was based were acquired slowly, after a great deal of experimentation. For this reason no one can expect to be able to understand or to utilize Mendel's law without a sincere effort. One may perhaps obtain a sketchy or restricted impression of its nature and significance and feel that he can then use the law to obtain a complete, easy, and foolproof solution of all the problems presented by the scientific breeding of dogs. *This is not the case.* Even though in color inheritance Mendel's law is the most important, the largest, the most universal, and the most predictable part of the story, and is therefore the key to those principles and procedures which lead to the most satisfactory and complete analysis now possible, it still is not everything.

A large number of inherited or transmitted influences which affect the *degree* of development or expression of the Mendelian genes can be identified. This type of minor influence is difficult or as yet impossible to analyze. Recognition of this fact will aid in evaluation of the application of Mendelism and will also avoid a great deal of wasteful expectation that all the most minute variations in coat color can be easily explained by Mendel's law.

PART TWO

Basic Coat-Color Genes of the Dog

CHAPTER III

Summary of Genes

EACH individual dog produces its own coat color. Each also determines what potentialities for coat color it can transmit through its sex cells to its progeny by the action of at least ten different genes, or gene pairs. As we shall see when the genes are separately discussed, there are various types of inter-action or of combined or related action between certain differ-ent gene pairs which affect the visible color type of the animal.

Like the other laboratory mammals, dogs appear to have two major types of pigment in their coats. One of these is yellow, the other dark (brown or black). The color varieties of dogs have to be formed by various genes controlling the amount, extent, and distribution of these pigments, both individually, in combination, or in competition with one another.

Pigment in the coat is formed by the interaction of two types of substances. One is relatively inert, widespread, and basic to the survival of the animal; this is called a *chromogen*, or "color base." Tyrosine is probably one of the most important sub-stances of this type in dogs. The other type of substance is known as an *enzyme*, or catalyst. The distribution of this sort of substance can be definitely varied, inhibited, or limited in extent. The interaction of these two types of substance results in the formation of granules of pigment, known as *melanin*.

Such pigment granules can be distributed in various amounts and patterns in either or both the outside layer (cortex) or the inner portion (medulla) of the hair. Variations in such processes produce different optical color effects resulting in the different color varieties of dogs.

All laboratory mammals form essentially the same kinds of pigment by basically the same types of chemical reactions. It is therefore interesting, and sometimes helpful in analysis, to compare the genetic behavior of color varieties which have very much the same appearance in the various types of mammals. An attempt will be made to make such comparisons when the different genes in dogs are considered.

The genes for coat color so far identified in the dog are as follows:

1. The *A* series. The genes which form a series of multiple alleles situated in the *A* locus of the chromosome influence the relative amounts and the location of dark (black or brown) pigment and of light (tan or yellow) pigment both in the individual hair and in the coat as a whole.

There are certainly three, probably four, and possibly five, alleles in this series. These form a series of patterns of different distributions of dark and light pigment.

The following list of alleles is in the order of epistatic position.

A^s—allows the distribution of *dark* pigment over the whole body surface. (Type—Newfoundland, Chesapeake Bay Retriever.)

a^y—restricts markedly the areas of *dark* pigment and in its most complete expression produces a clear sable or tan-colored dog. (Type—Basenji, Irish Terrier.)

a^t—produces the bicolor varieties (black-and-tan, liver-and-tan, etc.). All of these we shall describe as *tan-points*.[1] (Type—Welsh Terrier, Doberman Pinscher.)

The possible fourth allele, a^w, would probably fall in the series between A^s and a^y or between a^y and a^t. It would produce

---

[1] Dogs with this coat-color pattern have tan muzzles, feet, under surfaces of the tail, anal regions, etc. Because the dark part of the coat may be *either* black or liver, it is necessary to devise some term that will include *both* black and tan and liver and tan to show that they have the same genetic pattern gene. For this reason the term "tan-points" is used to include any manifestation of the pattern, no matter what the darker ground color may be.

the agouti or "wild-color" type of coat color with banded hairs seen in Norwegian Elkhounds, gray German Shepherds, Schnauzers, and certain wild Canidae. Until further data are obtained, it may be given the symbol a^w.

A number of publications on coat-color inheritance in dogs have hypothesized another allele, a^s or "saddle," in the A series. This pattern represents a type of bicolor (tan-point) distribution in which the dark pigment is restricted to a "saddle" and tan pigment covers the head, legs, and neck. It is seen in adult Airedale and Welsh Terriers and in Beagles. All of these breeds are distinctly bicolored at birth—dark pigmented with tan points. They rapidly grow lighter with age. This is also true of sables $a^y a^y$ or $a^y a^t$, which may begin as dark-pigmented animals and become progressively lighter. In sables there is no evidence of or claim for any relationship of a simple allelic nature between the darker and lighter types. At present it does not appear to the writer that the existence of the a^s allele is established. In both the sable and tan-point series the results of matings thus far recorded might be equally well explained by the theory of darkening or of lightening modifiers.

2. The B–b pair. This is one of the most completely analyzed pairs of coat-color genes in dogs.

The B gene produces *black* coat color, determined by a certain quantity and quality of dark-pigment granules in both the cortex and the medulla of the hair.

Its alternative and recessive form, b, allows the reduced degree of pigment formation observed in the liver (chocolate-brown) series of dogs.

3. The C series. There appear to be four well-recognized alleles in this series. By far the most frequent is C, the gene for full depth of pigmentation. This is the condition observed in the deep rich pigmentation of golden brindles, dark tans or reds, or deep, absolutely black or liver varieties. The second member of the series is c^{ch} or chinchilla. This gene has a distinctly greater effect in reducing the red-yellow pigment than it does

on the black pigment. For this reason, it becomes apparent chiefly, if not entirely, in light-pigmented areas. Norwegian Elkhounds and Schnauzers in which the dark pigment is still plentiful but the yellow has been so reduced as to be almost, if not entirely absent, from the coat are good examples.

The third member c^a, or complete albinism, resulting in a white animal with pink eyes, has been reported several times in dogs but is extremely rare and has never been used to develop a true albino variety or breed. Its existence is, therefore, a matter of theoretical, rather than of practical, interest.

There is a probable fourth allele, c^e, which lies between the chinchilla gene, c^{ch}, and complete albinism, c^a. This gene, called "extreme dilution," would perhaps account for extremely pale pigmentation affecting both *B* and *b* types. There might be a reduction of eye pigment to produce a ruby or reddish appearance.[2]

4. The *D–d* pair. Dogs with the *D* gene are called "intensely" or "densely" pigmented. Most dogs fall into this group. The gene *d* produces Maltese or blue dilution, most clearly recognized in short-haired dogs, such as blue Great Danes, blue Greyhounds, or blue Dobermans. It is also found in Chow Chows and in Poodles, among the longer-haired varieties.

5. The *E* series. The "black mask" pattern seen typically in Pugs and Norwegian Elkhounds as well as in a number of other breeds has been included as the top-ranking member of the *E* series of alleles. It is given the symbol E^m. Recent experiments suggest the following working hypothesis, to be tested by further experiments: black mask may be due to a sort of "super-extension" factor E^m epistatic to all other members of the *E*

---

[2] Burns (1952) lists five alleles. Her *C* is identical with mine. Her c^r is the same as c^{ch} and the c^a is identical. I have had no personal experience with "Cornaz" albinos (her c^b) and have considered that a "white coat with dark eyes" is due to $c^{ch}c^{ch}$ acting on a yellow background. Burns's c^d has therefore been omitted. Her interpretation may well prove to be correct. The extremely rare incidence of true pink-eyed albinos in dogs and cats is interesting.

series. This matter will be more completely discussed under the *E* locus when that topic is considered in more detail.

The next member of this series of alleles is usually considered to be *E*, which allows the formation of dark (black or brown) pigment evenly over the whole coat.

The bottom member of the series is *e*, in which no dark pigment can be formed in the hair, leaving it a clear and evenly distributed shade of red or yellow. Orange Pomeranians or Irish Setters are good examples.

Between *E* and *e* and often imperfectly and incompletely hypostatic to *E* but usually clearly epistatic to *e* lies e^{br} (brindle), which in the presence of a^y or a^t produces brindle dogs, with bands of dark pigment, more or less regular in outline or extent, on a background of tan or yellow. Scottish Terriers, Irish Wolfhounds, brindle Great Danes, and Boxers are examples.

6. The *G–g* pair. The gene *G* appears to be at least partly dominant and changes a puppy born of a uniform dark color in the direction of increasing grayness or paleness. The dog that remains without this lightening gene is *gg* in genetic make-up. The gene *G* is found in Kerry Blue Terriers, Poodles, and some other breeds. Its visible effect may easily be confused with that of *d*.

7. The *M–m* pair. Here we have a gene *M* which, acting upon uniform pigmentation (*mm*), produces merle (dapple) animals, often with areas of white on the coat. Occasionally one or both eyes are light blue or have heterochromida iridis (variegated or blotched pigmentation of the iris). In some breeds *MM* animals are badly deformed or handicapped. In Collies, for example, they are usually deaf and blind and are often sterile. Their coat color is usually white.

8. The *P–p* pair. This is a pair of genes dealing with a very rare variation known as "pink-eyed dilution." The most authentic and extensive description of its appearance is given by Pearson (1929), who, from data obtained from breeders, recorded its occurrence in Pekingese. Animals with the *P* gene are the ordinary color varieties of dogs. In *pp* animals, however, the *dark* pigment is greatly reduced wherever it appears, without reduction of the yellow or red. On the coat this makes a black dog a pale blue, which in some cases has been described as "lilac." Liver animals under the influence of *pp* become light yellowish fawn in color.

9. The *S* series. Here there seem to be at least four alleles. The highest in the series is *S*, responsible for a solid-colored coat, with no white or with very minute spots on the toes or chest. The next member of the series is s^i, known as "Irish" spotting because a similar pattern in the Irish rat was so described by Doncaster (1905). In this variety any one or more of the following areas may be white: muzzle, forehead, feet, tail tip, chest, belly, throat, or neck. This pattern is seen in perhaps its most typical form in the Basenji.

The third member in the series is s^p, responsible for piebald spotting. In this group one may find dogs varying from those which show a condition similar to Irish spotting in appearance to others which have only 15 to 20 per cent of their coat pigmented. Beagles show this range of expression of the piebald gene. An undoubted fourth and lowest member, s^w, causes extreme piebald spotting. It is found in breeds like the Bull Terrier and Sealyham Terrier, where an ear, eye, or tail patch occasionally appears in an otherwise white breed. There is bound to be some confusion in the visual classification of piebald dogs. If, however, one has access to the breeding records of individual animals and their progeny, one can usually determine the identification of the animals genetically, according to the four alleles listed above.

10. The *T–t* pair. Here the gene *T*, acting as a dominant, produces in white areas of the coat flecks or ticks of color, usually referred to as "ticking." In general, the behavior of the gene is fairly regular. There are, however, some modifications which affect its expression and which, upon further study and more extensive data, might necessitate the hypothesis of more than two alleles or some other genetic explanation. Spotted dogs without *T*—namely, *tt* in formula—should have clear-white, unticked areas.

The relation of roan to the *T* gene will be discussed later. By some, roan is considered to be due to a gene *R* dominant (in spotted animals) over clear-white spots.

It may be helpful to pick some types of dogs whose coat colors are produced by characteristic combinations of these genes. We shall suppose that each type is uniform for the genes listed. Since our hypothetical animals do not carry recessives of any of the listed genes, we can use only one symbol for each gene instead of two.

Newfoundland (black)	$A^s\ B\ C\ D\ E\ g\ m\ P\ S\ t$
Irish Water Spaniel (liver)	$A^s\ b\ C\ D\ E\ g\ m\ P\ S\ t$
Labrador Retriever (black)	$A^s\ B\ c^{ch}\ D\ E\ g\ m\ P\ S\ t$
Great Dane (blue)	$A^s\ B\ C\ d\ E\ g\ m\ P\ S\ t$
Great Dane (merle)	$A^s\ B\ C\ D\ E\ g\ M\ P\ S\ t$
Pekingese (pink-eyed dilute)	$A^s\ B\ C\ D\ E\ g\ m\ p\ S\ t$
Scottish Terrier (black)	$A^s\ B\ C\ D\ e^{br}\ g\ m\ P\ S\ t$
Kerry Blue Terrier (blue)	$A^s\ B\ C\ D\ E\ G\ m\ P\ S\ t$
Scottish Terrier (brindle)	$a^y\ B\ C\ D\ e^{br}\ g\ m\ P\ S\ t$
Irish Terrier (red)	$a^y\ B\ C\ D\ E\ g\ m\ P\ S\ t$
Basenji (red)	$a^y\ B\ C\ D\ E\ g\ m\ P\ s^i\ t$
Dachshund (black, tan-points)	$a^t\ B\ C\ D\ E\ g\ m\ P\ S\ t$
Dachshund (liver, tan-points)	$a^t\ b\ C\ D\ E\ g\ m\ P\ S\ t$
Beagle (tricolor–clear-white areas)	$a^t\ B\ C\ D\ E\ g\ m\ P\ s^p\ t$
Beagle (tricolor–ticked-white areas)	$a^t\ B\ C\ D\ E\ g\ m\ P\ s^p\ T$
Dalmatian (black spots)	$A^s\ B\ C\ D\ E\ g\ m\ P\ s^w\ T$
Norwegian Elkhound	$a^w\ B\ c^{ch}\ D\ E\ g\ m\ P\ S\ t$
Weimaraner	$A^s\ b\ C\ d\ E\ g\ m\ P\ S\ t$
Poodle (gray; born black)	$A^s\ B\ C\ D\ E\ G\ m\ P\ S\ t$
Poodle (blue; born blue)	$A^s\ B\ C\ d\ E\ g\ m\ P\ S\ t$
Poodle (light silver; born gray)	$A^s\ B\ C\ d\ E\ G\ m\ P\ S\ t$

These and other combinations will be more fully discussed under the particular breed headings.

It is evident that, because of the cumbersome and complex nature of the complete formulas, it will be more satisfactory to consider each locus separately and thereby gain some idea of the various ways in which each gene can express itself. Chapters IV–VIII will attempt to do this.

CHAPTER IV

Locus A

IT WILL be recalled that this locus contains three, and possibly four or five, alleles. The highest member of this series is A^s, which allows dark pigment to extend over the entire body surface.

Like other members of this allelic series, A^s is a pattern factor, the result of a sort of competition between a tendency to allow the spread of dark pigment and an opposing tendency expressed by a^y (sable, tan), a^t (tan-points), and possibly a^w (wild color, agouti) to *restrict* the spread of dark pigment and to replace it by hairs with little or no dark pigment.

Where two basic and independent pigment types are present, it is entirely logical to expect that there should be hereditary variations in their formation and distribution. This is true of the A series.

The existence of at least two such fundamentally distinct processes of pigmentation can be considered a basic phenomenon in mammals. In one of the pigment-forming processes, the brown (liver)-black pigments are involved. These two pigments, as we shall see when we consider the B locus, appear to be two distinct stages of an oxidation process involving a single enzyme. The yellow-tan hairs in the A series occur when the activity of this enzyme is prevented, greatly restricted, or decreased. Not only does the effect have a regional aspect, as is shown by differences between dorsal and ventral or central and peripheral areas of the body, it may also operate in individual hairs, producing a tip of one color (dark) and a base of the other (light) or a tip *and* a base of the same color type and a middle band of the other. Evidence from rodents (mice and rabbits) is clear that coat-color types involving either the regional or the individual hair pattern may be allelic to one another, e.g., agouti and black and tan.

We shall first consider the three well-established and alternative alleles in dogs, a^t, a^y, and A^s, and then comment on the evidence for the existence of a fourth allele, a^w (wild color or agouti).

a^t–*Tan-Points.* The so-called black-and-tan pattern is the lowest member of the multiple-allele series at the A locus. Because of the fact that this pattern occurs in liver (b) dogs as well as in black (B) and dilute (d) animals, the term "tan-points" is suggested, as mentioned earlier (p. 22).

In order to provide a basis for evaluating such a suggestion, we will discuss certain aspects of the varied expression of this pattern. Three main factors contribute to this variation: (1) the depth of the tan pigmentation; (2) the extent of the tan pigmentation; (3) the degree (if any) to which the existence of the tan areas in the pattern is discernible through the overlying dark pigment in A^s (self) [1] or a^y (tan) animals, carrying a^t as a recessive.

DEPTH OF TAN PIGMENTATION. In some individuals the tan pigmentation is so heavy and dark as to be easily confused with liver, from which, however, it is genetically completely distinct. It is often very difficult to discern the tan-point pattern in these animals because of the slight contrast between the usually light areas and the liver ground color of the coat in bb dogs.

From such extremes of dark pigmentation there extends a continuous series of lighter degrees of pigmentation in which the tan points are increasingly in contrast with the dark ground

---

[1] The term "self" is one commonly used by mammalian geneticists to denote a uniform solid-colored coat without white spots or pattern, as the case may be. It probably derives from usage established by breeders of pigeons or cavies who use it to classify animals at exhibitions.

color. There is, of course, a possibility that this may be due to a graded series of alleles with overlapping influence on the depth of pigment. Especially does it seem possible that the extremes on the lightly pigmented end of the series have a distinct and characteristic genetic basis of some sort. This will be discussed under the *C* locus.

At the Jackson Laboratory, for example, some Beagles, instead of the usual rich (intermediate) tan pigmentation, have had a dull, flat, light-buff color. Similar pigmentation has also been observed in tan-point hybrids derived from a cross between Pug, Dachshund and Beagle.

In both cases the light variation appears to be hypostatic, or recessive, since animals exhibiting it are obtained from parents both of which have intermediate or rich tan-points. The genetic explanation is being sought by further experiments.

When tan-point animals $a^t a^t e^{br} e^{br}$ with brindling in the areas usually clear tan have been produced in the Jackson Laboratory, the difficulty in detecting these areas has sometimes been great. Some puppies on casual inspection at birth seem to be solid black. In a certain light, however, a slightly brownish tinge can be observed on the areas which are tan in tan-point animals. Also there are usually some tan or yellow hairs under the tail. As the animals grow older, these areas become lighter until the tan-point pattern can be clearly recognized.

EXTENT OF TAN PIGMENTATION. The typical tan-point pattern includes areas of tan pigment on the sides of the muzzle, throat, and belly line, inside the ears, on the chest, over each eye, on all four feet and part of the legs, around the anus, and on the underside of the tail.

The area of tan may be so reduced in any or all of these regions as to make its detection difficult, and sometimes the spots actually disappear. Obviously, reduction in the pattern of dogs with *dark* tan results in a combination of effects producing the most obscure type of the pattern. It is probable that such animals would be classed as A^s or self, in which no tan areas are shown, unless a most careful examination or a genetic test to discover the tan-point gene was made.

Extension of the tan areas beyond the regions described above is frequently observed. The muzzle spots may extend upward to the forehead, head, and ears, and even well down on the neck. On the neck it may meet an extension of the foot spot up the legs and chest and onto the shoulder. A similar extension can occur up the hind legs, while the anal spot and tail streak may cover that whole region. At the extreme of extension, a dark saddle or broad dorsal streak is usually the only dark area remaining on an otherwise tan animal.

In some breeds, e.g., Dachshunds, the form of extension of tan which will characterize the adult animal is often essentially present at birth. In others, however, e.g., Airedale, Beagle, and Fox Terrier, the puppies are born with a relatively large area of dark pigment, and the tan extends progressively until the dog is full grown. In some individuals further reduction of the dark areas may be observed in old age.

This competition between tan and dark areas is also observed, with striking age effects, in the a^y and a^w members of this allelic series. A more complete discussion of this matter will be given under those alleles.

EXPRESSION OF TAN PIGMENTATION. There are some cases in which a self (A^s; black or liver) animal carrying a^t as a recessive shows a reddish or light-colored tinge in the dark pigment overlying the areas which, in an $a^t a^t$ animal would be tan. In some cases this phenomenon is visible only during a limited age range or in certain light. In others it may persist and be discernible by careful examination in any type of lighting.

Animals of this sort are often difficult to classify without a breeding test to determine the correct genetic composition of the individual. It is evident that errors in classification could easily lead to distorted and atypical ratios of color classes, which, in turn, might cause misinterpretation of the genetic nature of certain matings and their results. This problem is directly comparable to some of the complexities observed in the visible (phenotypic) relationships between a^y (tan) and a^t, which will be discussed in the section on the a^y gene.

At the Laboratory 61 litters from tan-point × tan-point ($a^t a^t$ × $a^t a^t$) matings in Fox Terriers, Dachshunds, Shetland Sheep Dogs, Beagles, or their hybrids have given a total of 292 puppies, *all tan-point* as expected. Data from cooperators' records are given in Table 1. From both groups of data it is clear that a^t is the bottom or lowest member of the allelic series.

Table 1. Tan-point × tan-point matings in several breeds

Breed	No. of litters	No. of pups	No. tan-point	No. other colors
Airedale Terriers	40	358	358	0
Beagles	147	775	747	28*
Dachshunds	25	118	117	1†
Wire-haired Fox Terriers	23	90	90	0
Total	235	1,341	1,312	29

* Of these, 25 were yellow and white. These could be (1) $a^t a^t$ animals with the extension factor E replaced by e (see E locus), (2) $a^t a^t$ tan-points with black areas coinciding with white spots and therefore invisible, or (3) a possible mutation to a^y in the sable-tan allele. This possibility is remote. The other 3 animals classed as "gray and white" were probably gray-sable and white and *were* due to such a mutation. They might, however, be "blue" dd animals erroneously described as gray.

† One puppy was described as "light brindle," probably a gray-sable or a^y mutant. It is not at all uncommon for a^y (sable-tan) animals carrying a^t as a recessive to appear grayish (tan) at birth.

a^y—*Tan or Sable.* Proceeding upward in the series, we have a^y, which is epistatic to a^t but hypostatic to A^s.

As in the tan-point series, the a^y character can manifest itself in a wide range of color types. These result from variation (a) in the depth of pigmentation and (b) in the extent of dark pigment both in the individual hairs and in regions of the coat.

DEPTH OF PIGMENTATION. Basenjis provide probably the best opportunity to establish a consistent "base line" for the appearance of the $a^y a^y$ type. They are short-haired, which eliminates variation in the depth of color effected by the distribution of pigment-forming activity or granules throughout a longer hair. Usually they are a uniform rich tan in the adult animal, although, as has been noted, newborn puppies may show considerable amounts of dark pigment.

In Collies the so-called sables are often a clear yellow-tan, and it is probable that these are $a^y a^y$ individuals.

Irish Terriers are also a clear uniform tan as adults. Newborn pups are considerably darker than adults, especially around the muzzle, foreface, and ears and along a middorsal line.

In Pugs, however, the fawn variety shows considerable variation. There are clear-red fawns which at maturity are similar to the Basenji tans. There are also silver fawns which appear almost gunmetal gray at birth and which, as adults, exhibit a dull, flat tan. It is probable that these are the result of the presence of the "chinchilla" type of pigment reduction, which will be considered under the *C* locus.

The depth of tan pigment in $a^y a^y$ dogs which are brindle or have banded hair also varies greatly. This is seen in Scottish Terriers, where the *light* pigment of brindle coats varies from a deep rich tan to such pale colors as to be almost white.

In Pugs the $a^y a^y$ deeply colored fawns almost always have black masks and considerable dark pigment in their coats. In Dachshunds the $a^y a^y$ pure-breeding reds, which do not carry a^t as a recessive, and even some $a^y a^t$ animals are clear and often *light* red with no black mask. As yet we have no record of the occurrence in Dachshunds of a silver-red with the peculiar dull, flat color of the silver-fawn Pug. It is interesting, therefore, to find in second-generation hybrids (F_2) between a red ($a^y a^t$) Dachshund bitch and a silver-fawn Pug dog some animals which are uniformly a very pale ivory or cream—almost white. These are probably the Dachshund type of $a^y a^y$ red with the gene (probably c^{ch}, see section on *C* locus) which produces the silvering in Pugs.

EXTENT OF PIGMENTATION. Most sable-tan dogs showing various areas of persistent dark pigment are probably $a^y a^t$ in genetic constitution. This is indicated by definite breeding tests of certain of these animals, by their frequency in breeds such as Dachshunds and Collies, where tan-point animals occur, and by their rarity, or absence, in breeds such as Irish Terriers and Basenjis, where most individuals are $a^y a^y$ in formula.

The amount and extent of dark pigment in such $a^y a^t$ animals vary tremendously. There may be a few scattered hairs with dark pigment that is scarcely recognizable unless a very careful and minute examination is made. There may be a dark stripe or body pigmentation even more extensive than the dark-pigmented areas of the least-developed types of the tan-point ($a^t a^t$) pattern itself. Between these extremes one finds a graded series of degrees of dark pigmentation. Such a situation has an analogue in the various types of sooty yellows and sables formed in mice by incomplete dominance of A^y, the highest member (yellow) of a multiple-allelic series. In this case dark pigment persists to a greater or lesser degree. In some of the $a^y a^t$ dogs areas of the coat superficially seeming to be dark-pigmented consist of hairs which upon close examination are found to be dark-tipped but light (red or tan) at the base.

From both the regional and the single-hair types of pigmentation, it seems probable that the a^y coat color is formed by a progressive restriction of the activity of the black-brown enzymes, either regionally or locally or both.

Laboratory data on crosses involving the a^y and a^t alleles are given in Table 2. It will be noted that the results observed and those expected on the allelic relationship hypothesized between a^y and a^t are in close agreement. Cooperators' records on Irish Terriers and on Basenjis are not very extensive, but their figures are given in Table 3.

In red Dachshunds and sable Collies there is no way to tell

Table 2. Various types of matings involving the a^y and a^t genes

Nature of mating	No. of litters	No. of pups	Observed		Expected	
			a^y, sable-tan	a^t, tan-point	a^y, sable-tan	a^t, tan-point
$a^y a^y \times a^y a^y$	10	56	56	0	56	0
$a^y a^y \times a^y a^t$	3	13	13	0	13	0
$a^y a^y \times a^t a^t$	7	45	45	0	45	0
$a^y a^t \times a^y a^t$	11	58	43	15	42.5	14.5
$a^y a^t \times a^t a^t$	20	79	34	45	39.5	39.5
Total	51	251				

Table 3. Matings of homozygous $a^y \times a^y$ animals in two breeds

Breed	Litters	No. of pups	a^y, tan-sable
Basenjis	1	5	5
Irish Terriers	12	66	66
Total	13	71	71

whether the animals used for breeding by the cooperators were $a^y a^y$ or $a^y a^t$ in genetic formula unless tan-point pups were actually produced. In each of these breeds, therefore, only two categories of matings are listed, namely red × red and red × tan-points. The numbers obtained are shown in Table 4.

A^s—*Self Color.* This gene is epistatic to both a^y and a^t. Its dominance over each of them is usually complete, as far as superficial examination is concerned. It has, however, been observed in some crosses that a peculiar reddish tinge appears through the black, sometimes only in certain lights, sometimes more generally. In no way is this to be confused with the relatively distinct contrasts in pigmentation seen in the incomplete coverage of the coat by tan areas in a^y or a^t animals. When A^s incompletely conceals the presence of a^y as a recessive, the indistinct reddish undertone is seen chiefly on the head, neck, sides, and legs and in adult animals may be clearly distinguished

Locus A

Table 4. Matings of a^y and a^t animals in two breeds

Breed	No. of litters	No. of pups	a^y, sable-tan	a^t, tan-point	% $a^t a^t$
			Red × red		
Dachshunds	78	420	352	68	16.3
Collies	249	1,872	1,658	145*	7.7
Total	327	2,292	2,010	213	
			Red × tan-point		
Dachshunds	84	419	271	148	35.3
Collies	67	686	452	200†	29.1
Total	151	1,105	723	348	

* In addition 17 were listed as black and white, 2 as blue-gray, and 50 as white. The black-and-white individuals may well have been tan-pointed with white areas making the tan invisible. The whites might either be extreme variants in the S series (see that locus) or the result of the merle gene in a homozygous condition (see M locus). The blue-gray individuals may either be indistinct variants from the typical merle or true dilute animals (see D locus). There is also, of course, the possibility of carelessness in describing a bicolor pattern as "black" or an animal with a small patch of pigment as "white."

† In addition, 9 black-and-white animals were listed, 1 blue-gray, and 24 white, possibly explicable as described under the preceding note.

as lighter than the rest of the animal. When a^t is the recessive pattern carried, the reddish tinge is found only where tan occurs in tan-point animals.

Many types of mating may involve A^s animals, and a number of these types have been recorded at the Laboratory. In some cases there are not many progeny, but the total evidence is consistent with our description of the relationship of this gene. The data obtained from the eleven types of matings recorded in Table 5 are consistent with, and in most cases close to, the expected results.

The relationship of A^s, a^y, and a^t is thus well established. It

Table 5. Various matings involving A^s, a^y, and a^t genes

Mating	No. of litters	No. of pups	A^s	a^y	a^t
$A^sA^s \times A^sA^s$	42	287	287	0	0
$A^sA^s \times A^sa^t$	3	20	20	0	0
$A^sA^s \times a^ya^y$	12	70	70	0	0
$A^sA^s \times a^ya^t$	3	14	14	0	0
$A^sA^s \times a^ta^t$	12	88	88	0	0
$A^sa^y \times A^sa^y$	2	12	9	3	0
$A^sa^y \times A^sa^t$	4	20	13	7	0
$A^sa^y \times a^ya^y$	2	11	4	7	0
$A^sa^t \times a^ya^y$	4	23	12	11	0
$A^sa^t \times a^ya^t$	3	12	5	6	1
$A^sa^t \times a^ta^t$	5	34	15	0	19
Total	92	591			

will later be seen that there is good evidence that this relationship is true, not only in the visibly different manifestations of these patterns, but also in the indistinguishable series of yellows produced by the *ee* genetic combination under which there is no dark pigment anywhere on the coat. This will be discussed under the *E* locus.

a^w—*Wild Color.* On the basis of data so far obtained at the Laboratory, it cannot be said that this gene has been definitely established as an allele in the *A* locus, or even as an independent unit. There is, however, distinct probability that such an allele exists. Certain hairs of the gray wolf and of the coyote are banded and not uniform in color throughout, and a banded type of hair makes up the coats of a vast majority of the wild rodents. (The name of one of the latter has been used by geneticists as the type name.) On the other hand, in the range of a^ya^y and a^ya^t types individuals occur with hairs clearly banded in restricted areas. These areas are usually along the border of solid dark (black or brown) areas. In some of the sable longhaired types (German Shepherds) they may be distributed on the neck, flanks, and legs. Then, too, in the brindle types to be considered under the E locus, banded hairs are found. In these animals the a^y and a^t alleles may well contribute to the general similarity to the wild-coat color pattern which results.

Two litters out of a Norwegian Elkhound (a^wa^w) by a tan (a^ya^y) sire at the Jackson Laboratory are of interest. All 16 pups were a very dark agouti at birth. Their coats were distinctly dark with banded hairs. They were, however, lighter than purebred Elkhound pups, which at birth are black, or black with scattered lighter hairs. These results, coupled with the well-known fact that a^ya^y, a^ya^t, and a^ta^t animals usually become lighter as they grow older, suggest that the wild-coat color, or gray-sable, seen in gray wolves, Elkhounds, certain German Shepherds, and some other breeds is due to an allele at the A locus.

Unfortunately the Elkhound at the Laboratory recorded as the dam of the 16 puppies mentioned above has abnormally large nipples, which prevent successful nursing by her pups. No foster mother could be obtained for the first litter. Of the second litter, 3 have been successfully reared, and it is hoped that if they mature their coat color will give definite evidence for or against the working hypothesis that the gene A is an allele in this series.[2]

It is probable that further properly planned crosses of such breeds as Schnauzers or Norwegian Elkhounds with the right genetic combination of A^s or a^t animals will be needed for complete clarification of what still seems to be a somewhat obscure classification of the so-called wild-coat color of dogs. Until that time, it seems best to recognize the probability that an allele a^w in the A locus exists and to admit the need for further evidence.

---

[2] At present the body color of these puppies is very like that of many a^y sable-tan animals. The remaining dark hairs appear to be banded, but the color is predominantly yellow. This suggests that a^w may be hypostatic to a^y. Such a relationship would be consistent with the hypostasis of the tan-point pattern, which is even more pigmented than is wild color.

CHAPTER V

Locus *B*, Locus *C*, and Locus *D*

THE *B* LOCUS

THIS is a well-established and clear locus, in which, at present, only two genes have been identified. The dominant member of the pair, *B*, causes the dark pigment at any location to be black. The recessive member, *b*, produces liver (brown) pigment. Dominance apparently is complete and segregation clear-cut and definite.

Matings at the Laboratory involving this pair of genes are recorded in Table 6. The correspondence between observation

Table 6. Various matings involving the *B* and *b* genes

Nature of mating	No. of litters	No. of pups	Observed		Expected	
			Black	Liver	Black	Liver
BB × *BB*	161	814	814	0	814	0
BB × *Bb*	11	92	92	0	92	0
BB × *bb*	23	159	159	0	159	0
Bb × *Bb*	15	91	70	21	69	23
Bb × *bb*	4	30	16	14	15	15
bb × *bb*	23	152	0	152	0	152
Total	237	1,338				

and expectation is clear. Supplementing these primary data are some obtained from cooperators' blanks. In Table 7 we present the results of matings of liver (brown) *bb* animals *inter se*. In these matings the 2 black puppies, which are exceptions, are probably due to mismating or to an error in the records.

Table 7. Matings of *b* × *b* animals in various breeds

Breed	No. of litters	No. of pups	Black	(Brown) liver
Chesapeake Bay Retrievers	23	189	0	189
Dalmatians	1	8	0	8
Standard Poodles	3	24	0	24
Miniature Poodles	1	3	0	3
Weimaraners	16	125	0	125
Doberman Pinschers	5	35	2	33
Total	49	384	2	382

In matings between blacks and browns there is no way in which one can be certain whether litters containing only blacks are from matings of a black *BB* with a *bb* liver or of a black *Bb* with a *bb* liver. The latter type of mating theoretically should produce a ratio of one black to one brown. In small or medium-sized litters, however, the departure from the expected ratio may be considerable. It is necessary, therefore, to tabulate all black × brown matings from cooperators' blanks, as recorded in Table 8. The distinct excess of black over brown is to be expected, if *B* is dominant and *b* recessive. Supporting evidence is also obtained from matings of black × black within these same breeds.

Table 8. Matings of *B* × *b* animals in various breeds

Breed	No. of litters	No. of pups	Black	(Brown) liver
Dalmatians	10	66	60	6
Standard Poodles	8	47*	27	19
Miniature Poodles	3	16	10	6
Doberman Pinschers	19	142	113	29
Total	40	271	210	60

* There was, in addition to the blacks and browns, one white pup. Since the nose color is not recorded, it is impossible to classify it as *B* or *b*.

Table 9. Matings of B × B animals in various breeds

Breed	No. of litters	No. of pups	Black	(Brown) liver	Other colors, unclassified
Dalmatians	56	411	400	11	0
Standard Poodles	17	122	92	13	17*
Miniature Poodles	17	64	54	6	4*
Doberman Pinschers	49	410	383	27	0
Total	139	1,007	929	57	21

* Poodles occur in various shades of apricot and cream and also in white. All these varieties are reasonably common. In Dalmatians, on the other hand, any color except black or liver is very rare. The same is true of Dobermans.

The variation in shades of brown (liver) from very dark to extremely light appears to be due to genetic factors other than changes in the b gene itself. The B–b pair remains, therefore, well established and a clear-cut example of simple Mendelian inheritance.

THE C LOCUS

Three and probably four alleles appear to be established at this locus.

c^a—*Complete Albinism.* Complete albinism (c^a, the most hypostatic of the alleles at the C locus) is extremely rare in dogs. A dog which is $c^a c^a$ in genetic constitution shows *no* pigment in hair, skin, or eyes. The hair is white, the skin a very light pink, and the eyes, where the blood vessels on the retina provide the only color, are *pink* or *red*.

Although a great rarity in dogs (Canidae) or cats (Felidae), many complete albinos have occurred in the carnivores, such as raccoons and ferrets. They are also widespread among rodents; mice, rats, guinea pigs, rabbits, squirrels, and woodchucks are examples.

In dogs and cats Burns (1952) has suggested that the true albino may be blue-eyed. This, however, seems unlikely. In cats blue eyes, either unilateral or bilateral, appear as a part of a dominant white (*W*), akin to some of the spotting genes and very definitely a pattern. This is shown by the occurrence of pigment spots on the forehead in some heterozygous *Ww* animals. The data of Pearson and Usher (1929), cited by Burns, show that fully pigmented coats occurred from crosses of two types of blue-eyed "albinotic" Pekingese ("dondo" × "cornaz"). This suggests that blue eyes are not necessarily characteristic of true albinos. If the parents had been true albinos, no fully pigmented young could have been produced. In rabbits, also, blue-eyed whites occur, not as a part of the albino series, but in a certain type of extreme spotting.

The author has never observed the pink-eyed white variation in dogs, but he has an apparently authentic record of its occurrence in a litter of Pekingese which were also pink-nosed. This litter was reported by a breeder in Kenosha, Wisconsin, in 1949. Whitney (1947) has also mentioned evidence of the appearance of this variation.

Care should be taken, therefore, not to confuse true albino dogs with white-coated dogs having blue or dark eyes until further data showing them to be genetically similar are obtained.[1] In the blue-eyed type some kind of spotting or of merle pattern is almost certainly involved in most cases. The genes causing dark-eyed whites will be discussed in the following paragraphs.

c^{ch}—*Chinchilla*. This gene is the next member of the trio of well-established alleles, being epistatic to c^a and hypostatic to *C* (full pigmentation) which will be considered later.

Schnauzers and Norwegian Elkhounds, in which the black pigment is little, if any, reduced and yellow-tan has become a light cream or almost white, are good examples of the effect

---

[1] The fact that the retina of certain blue-eyed white dogs and cats is so lightly pigmented that the pink tint of retinal blood vessels is seen through the pupil adds to the possibility of confusion. The iris, however, in true albinos also appears pink and not blue. That superficially similar coat colors can be considered homologous in different species is very doubtful.

of this gene. Its results are also seen in the light-gray brindle Scottish or Cairn Terrier and probably in a number of other breeds.

In the individual hairs of chinchilla animals pigment is reduced by the formation of both fewer and smaller pigment granules. Red or yellow is affected before and more extensively than black or brown. In some cases the yellow or red areas are light cream-colored or actually white in appearance because practically all the granules are absent.

The c^{ch} gene has little or no visible influence in solid-colored black dogs, and it is quite possible that a number of black breeds may really be $c^{ch}c^{ch}$ rather than CC in constitution. Liver (brown) animals, however, may show a direct influence of c^{ch}, and in this respect resemble brown (b) rodents of a similar genetic type. The reduced degree of dark-pigment formation observed in the bb as opposed to the BB type allows the still further step toward less pigment (c^{ch}) to express itself visibly.

In such a breed as Chesapeake Bay Retrievers, one finds a wide range of variation in depth of pigmentation. It may well be that the deep liver specimens are CC, the intermediate degrees Cc^{ch}, and the lightest shades $c^{ch}c^{ch}$ in genetic formula. Since, however, there seems to be a continuous graded series, the identification of the genetic types in an animal like the dog would be statistically difficult.

The c^{ch} gene seems to give the clearest effect in the tan and the yellow varieties of dogs. Among the tan-yellow types, those which result from action of the e gene and which also possess the gene G (pp. 69–70) seem to be the ones which show this reduction most markedly, although when more combinations of $a^{y}a^{y}$ with $c^{ch}c^{ch}$ have been obtained, this statement may need to be modified.

There is reasonably good evidence also that the reds and fawns of Greyhounds and the reds of Cockers can be lightened to light-yellow cream, or perhaps even white, by the action of c^{ch}, whether they are the rare a^{y} or the common e reds and fawns.

A gene for extreme dilution (c^e) may also be present as another allele in this series at the C locus. A gene of this type might act upon tan-yellow dogs to produce pups which would be white at birth and which would remain so nearly white during increasing maturity that traces of very light-yellow pigment would be hard to find and/or would be definitely localized. Among the breeds which seem to offer evidence for the existence of the c^e gene are West Highland White Terriers. Occasionally pale-yellow pigment may be found in scattered hairs, or at the base of the hair, in certain areas, notably along the middorsal surface.

There are a number of reasons why the definite identification of a c^e allele would be very difficult. First, its probable effect, if any, on the reduction of dark (black or brown) pigment is unknown. If this reduction effect is slight—and the rarity of color varieties in dogs similar in appearance to the known extreme-dilute types of other laboratory animals may indicate that this is the case—analysis of it would have to be largely confined to the red-yellow series. In support of the statement that dogs showing the greatly reduced dark pigmentation to be expected of extreme dilutes are rare is the fact that the only white dogs with a relatively nonlocalized distribution of scattered dark hairs are occasional sled dogs of the Malamute or Siberian Husky type. These are *not* to be confused with the light silver-gray types, unless they represent very marked reductions in pigmentation which might occur as an extreme variation of that pattern.

Second, the fact that brown-nosed (bb) reds or yellows are not so favored in many breeds as are black-nosed (BB or Bb) animals would mean that very light, extreme-dilute browns would not be saved for breeding. Poodles may be an exception; this will be discussed in the section on this breed.

Third, the degree of pigmentation seen in the palest types of $c^{ch}c^{ch}$ reduction may well overlap and be indistinguishable by eye from the darker specimens of extreme dilutes. This would make it difficult to separate the two alleles in tabulating ratios of their possible incidence.

C—*Full Pigmentation*. This condition is found in all deep-red or tan breeds, such as Miniature Pinschers and Irish Setters. It is probably also characteristic of many of the dark-liver breeds, such as Irish Water Spaniels, and of some black breeds, such as Newfoundlands and Schipperkes.

Since blacks, however, do not visibly show the effects of c^{ch}, the presence of that gene may be masked. An example of this sort may be Labrador Retrievers, in which, when the mutation to yellow occurred, the yellows were pale, not deep red.

At the Jackson Laboratory fully pigmented dogs have produced only fully pigmented offspring in most matings. Among the breeds which have seemed to be uniform in this respect are Basenjis, Wire-haired Fox Terriers, Irish Setters, Doberman Pinschers, Great Danes, and Shetland Sheep Dogs.

Evidence of the presence of the c^{ch} gene carried as a recessive by fully pigmented C animals has been obtained in such breeds as Chows, Chihuahuas, Cockers, and Scottish Terriers.

THE *D* LOCUS

This contains another well-established pair of genes producing alternative characters. The dominant member, *D*, causes intense pigmentation, seen in deep rich-colored dogs such as Schipperkes, Irish Water Spaniels, and Irish Setters. The recessive member, *d*, resulting in blue or Maltese dilution, is seen in Great Danes, Greyhounds, Poodles, Chows, and Weimaraners.

Between the two members of the pair, clear-cut segregation is found with no confusing intermediates. The only difficulties in classifying *D* and *d* individuals are encountered in the light tan-yellow types of coat color. Here the fact that c^{ch} or *G* also decreases the amount of pigment creates a situation in which it is not always easy to determine by inspection whether c^{ch}, *d*, or a combination of the two genes has caused the observed coat color.

One characteristic frequently seen in *dd* animals is the peculiar flat, dull quality of the color, whether it is dark or light. This appearance has been described as "silvery," a quality very prominent in the dilution of liver, as seen in Weimaraners, which are *bbdd* in formula. An exact replica of the Weimaraner color has been deliberately created at the Jackson Laboratory by this combination of genes in a second-generation hybrid from a cross between a liver *bbDD* pointer dog and a blue *BBdd* Great Dane bitch. The F_1 generation consisted of black *BbDd* animals, and the dilute liver *bbdd* recombination expected in one of 16 F_2 animals was recovered.

The results of crosses involving the *Dd* pair of alleles, as recorded at the Laboratory, are shown in Table 10.

Table 10. Various matings involving *D* and *d* genes

Nature of mating	No. of litters	No. of pups	Observed		Expected	
			D, intense	*d*, dilute	*D*, intense	*d*, dilute
DD × *DD*	201	1,121	1,121	0	1,121	0
DD × *Dd*	17	122	122	0	122	0
DD × *dd*	4	28	28	0	28	0
Dd × *Dd*	11	51	38	13	39	13
Dd × *dd*	1	2	1	1	1	1
dd × *dd*	2	6	0	6	0	6
Total	236	1,330				

Because the breeding of intense (*D*) and dilute (*d*) varieties is not regarded with equal enthusiasm in most dog breeds, the blanks received from cooperators add very little data. Only four breeds contributed information of value. Table 11 summarizes the data. The cooperators' records are consistent with the direct genetic evidence obtained at the Laboratory and establish clearly the *D–d* pair of alleles.

Table 11. Matings of $D \times D$, $D \times d$, and $d \times d$ animals
in various breeds

Breed	Nature of mating	No. of litters	No. of pups	D	d
Chows	$D \times D$	43	216	183	33
Dobermans	$D \times D$	77	610	605	5
Great Danes	$D \times D$	39	368	366	2
Total		159	1,194	1,154	40
Chows	$D \times d$	5	19	17	2
Great Danes	$D \times d$	2	24	23	1
Total		7	43	40	3
Chows	$d \times d$	1	4	0	4
Great Danes	$d \times d$	3	25	0	25
Weimaraners	$d \times d$	16	125	0	125
Total		20	154	0	154

CHAPTER VI

Locus E

THIS is one of the most interesting loci in dogs. At it are found four alleles: E^m, superextension mask; E, extension, or solid coat color; e^{br}, partial extension, or brindle; and e, restriction, or red-yellow.

This is a difficult group of alleles to define and to understand because of the following complications: (1) there is considerable variation in the expression of, and in interrelations between, the E and the e^{br} alleles; (2) the interrelationships existing between coat colors produced by the A series and the E series of alleles are varied and at times confusing as to identification of color types. It will be well, therefore, to consider first those cases in which the visible influence of genes in the E locus is relatively simple and then take up the more complex examples.

e—*Red-yellow Pigmentation*. Dogs which are ee in formula exhibit a wide range of pigment quality and depth. They may be a deep rich mahogany as in Irish Setters, an intermediate red-fawn as in Golden Retrievers, or a lemon- or orange-yellow as in Pointers or English Setters. In all cases no dark (black or liver) pigment is found in the hair. This sharply differentiates the ee animals from sable or sooty animals produced by the action of $a^y a^y$. This matter will be more completely discussed when the relationship between the A and E loci are considered.

At the Laboratory 14 litters from $ee \times ee$ matings have produced 103 pups. All are red or yellow, as the hypothesis requires.

From cooperators' blanks, data on two all-red breeds have been tabulated. These are shown in Table 12.

Fig. 3. The extension (*E*) series in Great Danes. In the left-hand column are E^m (black-masked animals) of the formulas $E^m A^s$ (top row); $E^m{}_e{}^{br}a^y$, black-masked brindle (middle row); and $E^m a^y$, black-masked tan (bottom row). In the right-hand column are $E A^s$ (top row), $e^{br}a^y$ (middle row), and $E a^y$ tan (bottom row).

Table 12. Matings of $e \times e$ animals in two breeds

Breed	No. of litters	Total pups	Yellow or red
Golden Retrievers	44	377	377
Irish Setters	129	1,197	1,196*
Total	173	1,574	1,573

* One puppy is recorded as "black," which probably means that a mutation from e to E occurred. There seems to be little chance of error in this record, for the exceptional animal occurred in a litter of 11, produced by red parents. It was a typical Irish Setter in conformation, as were its 10 sibs.

Interrelationship of e *and* A. Since the interrelations between the genes at the E locus and those at the A locus produce various interesting coat-color modifications, this topic will be discussed under each allele of the E series.

Red or yellow animals which are eeA^sA^s, eeA^sa^y, eeA^sa^t, or eea^ya^y are probably evenly pigmented throughout the coat. In other words, there is no evidence of a pattern making some areas of the coat dark and others light. On the other hand, eea^ta^t animals in which the red-yellow pigment is sufficiently dark may show lighter-yellow points against the darker-red background. This pattern may persist throughout life, or it may disappear with increasing age.

Identification of a^ta^tee animals has been established. An example is ♀ 0211, an orange-and-white English Setter, which, when crossed with a black-tan-point Doberman, a^ta^tEE, No. 0269, gave 11 puppies, all black-and-tan a^ta^tEe in formula. Another example is ♀ 0690, a red Cocker derived from two tan-point a^ta^t parents, which produced 8 tan-point pups by a Beagle sire.

Similarly, individuals whose genetic formulas are A^sA^see have been identified. Examples are shown in Table 13.

The genetic analysis of the two types of red-yellow (a^ya^yEE and A^sA^see) in combination with one another has not yet been

Table 13. Results of mating male black-tan-point Doberman
($a^t a^t EE$), No. 0269, with various $A^s A^s ee$ females

♀ Parent	Kennel No.	Offspring
Cream Standard Poodle	0228	8 black or liver pups
Irish Setter	0230	8 black*
Yellow-red Irish Setter × Poodle F_1	307	10 black

* A number of these black pups showed a distinct reddish tinge, or sheen, in certain lights. While this is a difficult character to measure, it is very definitely present and produces a coat color which, under certain conditions, appears very different from the deep uniform black of such breeds as Schipperkes, Labrador Retrievers, or Newfoundlands. It seems that while the $A^s A^s ee$ × $a^t a^t EE$ cross produces a "synthetic" black $A^s a^t Ee$ by the simultaneous presence of A^s and E, the hypostatic alleles a^t and e result in enough reduction of dark pigment to allow a suggestion of red in the dark-pigmented hairs.

carried to a point where sufficient experimental data are available for final conclusions. Some results have been obtained, however, which support the hypothesis that the interaction of e and a^y is not complicated and that it will follow the genetic pattern set by the other alleles at the A locus.

One such cross is that of two matings of ♀ 0211, the orange-and-white English Setter already shown to be $a^t a^t ee$ in constitution, with a tan Basenji hybrid $a^y a^y EE$. Since a^y is epistatic to a^t, and since the combination $a^y a^t$ leaves little, if any, dark pigment on which E can act, the expectation is that sable or tan puppies will be produced. Actually the matings gave 22 sable or tan pups (a^y).

Three litters were obtained from tan-yellow $a^y a^y EE$ bitches (one Basenji, one Irish Terrier) with red or buff Cocker Spaniel dogs $A^s a^t ee$ in formula. The expectation is equal numbers of black $A^s a^y Ee$ and of red-tan or sable $a^y a^t Ee$. The numbers obtained were 10 black and 7 red, tan, or sable, a very close approximation.

From cooperators' and kennel club records have come some additional data which should be discussed. First may be cited the results obtained in Pointers. These data were published by the author many years ago (1914). In this breed yellow or lemon is probably the result of the *ee* combination. The figures obtained in various types of matings are shown in Table 14.

Table 14. Matings involving *E* and *e* genes in Pointers

Mating	Total pups	*E*	*e*
Black or liver *E* × *E* black or liver	535	505	30
Black or liver *E* × *e* yellow or lemon	82	62	20
Yellow or lemon *e* × *e* yellow or lemon	4	0	4

These figures support the hypothesis that the *E* and *e* alleles behave as alternative mendelian genes.

Just as tan-red animals produced by the combination $a^y a^y$ or $a^y a^t$ can vary greatly in depth of color, so red-yellow individuals that are *ee* in formula show all grades of pigmentation from deep mahogany to very light cream, which may at birth appear white.

It seems probable that the very dark and medium reds are those which have *C*, the gene for "full" pigmentation, and that in breeds like Irish Setters the selection of modifying genes has been long continued. This has resulted in the accumulation of a maximum amount of the red pigment.

In the other direction, it is likely that pale-yellow or cream coat color is produced by the interaction of the c^{ch} or c^e genes with the *ee* combination. There may also be, as perhaps in very pale silver-buff Cocker Spaniels, selection for reduced pigmentation, produced by modifying genes as yet unidentified individually.

e^{br}—*Brindle*. This is one of the more complex coat colors. Because of the wide variation in its expression and its interaction with genes at the *A* locus, it has been difficult to analyze genetically. In some breeds, such as Boston Terriers, the brindle pattern is confined to very dark "seal-brown stripes," fusing into and easily confused with, black. In these animals banding of the individual hair is usually absent or difficult to detect. Animals of this type are often erroneously described as "black." In Scottish Terriers, on the other hand, brindling is usually accompanied by banded hairs. It is probable that these are due to the activity of genes in the *A* locus and are not a product of the brindle gene alone. Evidence of this is obtained when the outer coat of a Scottish Terrier is removed by close clipping or plucking, leaving only the soft undercoat, where the stripes of light and dark pigment are clearly visible without any evidence of banding in the individual hair.

Brindle also occurs commonly in such other breeds as Greyhounds, Deerhounds, Irish Wolfhounds, Bulldogs, Boxers, and Great Danes. However, the areas of the brindle pattern may be so small that the dark area is entirely overlooked and the individual carelessly classified as fawn.

There is little doubt that e^{br} is epistatic to e. A cross between an English Setter (ee) and a brindle Greyhound ($e^{br}e$) resulted in 4 brindle ($e^{br}e$) and 3 yellow (ee) pups.

The genetic nature of fawn in Great Danes and in Boxers is still not entirely proved, because of lack of data on the cross-breeding of these animals with known genetic types of red-yellow-tan. At first the fawn of Boxers and Great Danes seemed to be due to the action of ee. Recently, however, a red hybrid bitch, known to be A^sA^see in formula, was crossed with a fawn Boxer dog and gave 7 pups, all black. This would definitely put the Boxer fawn in the a^ya^y series.

If this is the case, the relation to brindle in this breed would be as follows:

The fawn Boxer would be a^ya^yEE.

Brindles would be of two types: (1) $a^ya^yEe^{br}$, (2) $a^ya^ye^{br}e^{br}$.

The evidence that the two types of brindle can be produced will be given when the relationships of the *A* and *E* loci are discussed. For the present, we shall remove the fawn Boxers and

Danes from the *ee* series to the $a^y a^y$ group and shall consider the E and e^{br} relationship to be that described above.

Results of mating fawn × fawn Boxers have already been reported. They showed 808 fawn-tan and 2 brindle pups. Data published on Great Danes by Little and Jones (1919) showed 60 fawn and 4 brindles from fawn × fawn mating. When one realizes that some individuals that are genetically e^{br} may show so little brindling as to be classed as fawns, the exceptions can easily be explained.

In Great Danes, matings between brindles and fawns are more numerous than matings between fawns. Of a total of 476 pups produced by brindle × fawn matings, 291 were recorded as brindle, 185 as fawn. This result is consistent with the hypothesis that brindle is epistatic. In this breed, matings of brindle × brindle occur as frequently, if not more often, than do matings of brindle × fawn. Since homozygous $e^{br}e^{br}$ animals can be produced from the former type of mating and not from the latter, the general population of Great Danes will contain an appreciable number of homozygous brindles, which, when crossed with fawns, could account for the excess of brindles, 291–185, produced by such matings.

In Boxers, on the other hand, cooperators' blanks have recorded only 6 litters from brindle × brindle matings, while 67 crosses between brindle and fawn have been reported. This means that the vast majority of brindle Boxers are heterozygous and therefore that in crosses between brindles and fawns slightly more than 50 per cent brindles would be expected. Actually 216 brindles and 204 fawns were recorded, an excess of only 1.4 per cent. The six matings of brindle × brindle Boxers gave 33 brindles to 6 fawns, which is very close to expectations. Great Danes, in a total of 566 pups from brindle × brindle matings, produced 493 brindles and 73 fawns, which is also very close to what would be expected if the $a^y a^y$–$E e^{br}$ relationship existed.

Only a small number of crosses between e^{br} (brindle) and E

(solid-colored; black or liver) dogs have been made at the Laboratory, but these show clearly that the effect of the *A* series on the order of this epistasis in the *E*, *e^{br}*, *e* series is a very definite one. This will be discussed shortly. In the meantime it is sufficient to say that *A^s* must be present in at least a heterozygous condition for a solid-black coat to be clearly and completely epistatic to a brindle one. Such a condition was demonstrated when a black Schipperke *EEA^sA^s* was crossed with a brindle-hybrid bitch. All 4 pups were clear black.

Warren (1927) in a paper on coat-color inheritance in Greyhounds, where both black and brindle are standard colors, reported 68 blacks and 15 brindles from black × black matings. Little and Jones (*loc. cit.*) in matings of various types of black *inter se* obtained 287 blacks and 29 brindles. Also from black or dilute-black × fawn Great Danes, 21 out of 35 pups were brindle, undoubtedly produced by *Ee^{br}* black or blue parents.

Since, however, the darkest brindles, and those with the smallest area of brindle development, may *appear* black in certain breeds, the statement that black is epistatic to brindle must be carefully qualified and restricted according to the breed or cross in question.

Boston Terriers and Scottish Terriers are breeds in which the relationship of black and brindle is not simple or clear-cut. In both breeds modifiers that affect the *extent* of the pattern are present. In Boston Terriers the lighter stripes in the brindling may be so narrowed and so darkened by the action of modifiers that they become indistinguishable from the darker areas. In both breeds the depth and distinctness of the pattern may vary with age. This series of complications, alone or in combination, produces unavoidable confusion and requires skill and patience in its analysis.

Table 15 based on cooperators' records will show the situation in puppies at, or soon after, birth. From these figures it is very clear that the relationship between black and brindle in these breeds is not so simple as it is in Greyhounds and Great

Table *15*. Results of matings involving brindle and black in
Boston and Scottish Terriers

Type of cross	Boston Terriers Total				Scottish Terriers Total			
	Matings	pups	Black	Brindle	Matings	pups	Black	Brindle
Brindle × brindle	59	229	16	213	16	79	25	54
Brindle × black	31	122	56	66	24	115	67	48
Black × black	4	12	10	2	55	270	249	21
Total	94	363	82	281	95	464	341	123

Danes. This statement applies to black and brindle as visible coat colors only. The genetic behavior of E and e^{br} is undoubtedly orthodox and similar to that in other breeds.

E—*Extension of Dark Pigment*. At the E locus there are apparently three well-established alleles: E, e^{br}, and e. And all available evidence indicates that they segregate clearly and have predictable and definite effects, in the order of epistasis listed.

Reserving the further consideration of e^{br} (brindle) to its relationships to the A series, we now focus attention on the relation of the E and e alleles, with which most of the matings at the Laboratory have been concerned. The data are given in Table 16.

E^m—*Superextension of Dark Pigment* (*mask*). The clearest and most typical expression of this pattern occurs in a fawn $(a^y a^y)$ breed like the Pug. Here the black pigment covers the face and ears and may also be scattered along the middorsal line.

Evidence that if the gene for this pattern belongs in the E series it is epistatic to E is clearly provided by a cross between a tan Dachshund $a^y a^t EE$ and a masked fawn Pug $a^y a^y E^m E^m$. All 8 pups produced by this mating were tan (fawn) with black

Table 16. Matings involving E and e genes

Type of cross	Total litters	Total pups	E	e^{br}	e
$EE \times EE$	151	758	758	0	0
$EE \times Ee$	4	20	20	0	0
$EE \times ee$	33	236	236	0	0
$Ee \times Ee$	4	23	19	0	4
$Ee \times ee$	12	67	32	0	35
$ee \times ee$	14	103	0	0	103
Total	218	1,207			

masks. If the black mask had been due to an allele hypostatic to E, the F_1 would all have been clear tan (fawn).

When E^m replaces E in tan-pointed ($a^t a^t$) animals, the tan areas disappear from the head. Such dogs are frequently seen in black-and-tan Alsatians. A Newfoundland with tan-points, recently observed, had no tan on the head, suggesting that E^m was present. Wild-colored breeds ($a^w a^w$) such as Norwegian Elkhounds are probably $E^m E^m$ in formula, for the black mask is regularly present.

If the gene E^m is an allele of e^{br} (brindle), one may expect black-masked brindles, but it should not be possible to "fix" the mask as a homozygous character since if a second E^m appears the gene e^{br} is, by the theory of alleles, excluded.

When the gene E^m is found in combination with A^s, a "super black" animal should be produced. There should be no visible distinction between the black of the mask and that of the rest of the coat. When $E^m E^m A^s A^s$ blacks are used in crosses, the a^y, a^w, or a^t animals obtained as their descendants should show the mask in contrast with the lighter surrounding or adjacent areas.

Relation of E *and* e^{br} *to the* A *Series.* Dogs which are EA^s in constitution are black or liver, with no tan points and no brindle pattern. They may be deep and uniform in color and they are probably always so if they are EEA^sA^s in constitution.

Fig. 4. Combination effects of the A series and E^m (black-mask) genes in German Shepherds. Top row from left to right: $E^m A^s$, $E A^s$, and $E^m a^y$ in the clear-tan phase. Middle row, from left to right: $E a^y$ in clear-tan type; $E^m a^y$ in the dark-sable phase; and $E a^y$ in the same phase. Bottom row, left to right: $E^m a^t$ and $E a^t$.

A heterozygous condition of Ee and/or of A^sa^y or possibly A^sa^t may allow a reddish sheen in the coat or slightly less dark pigment at the points.

When an animal is a^ya^y (sable-tan), the E gene has little dark pigment on which to work. Sable dogs probably represent a conflict between E, which attempts to form or extend dark pigment in the coat, and a^y, which attempts to restrict or inhibit its formation. It is possible, therefore, to have a^ya^yEE animals that are all shades of sable including clear tan and that, when crossed with A^sA^see reds or yellows, recreate a solid dark-colored dog, A^sa^yEe.

There is considerable evidence that these two types of red-yellow coat colors (a^y and e) can coexist in certain breeds. There is no way to check the *proportions* in which they occur, but the qualitative evidence for their coexistence may be sought for, recognized, and presented.

Cooperators' blanks are useful in providing evidence of the coexistence of the two categories of reds within a breed. Most of the data bearing on this point has been obtained from the breeders of Cocker Spaniels. This is in keeping with the experience of the American Kennel Club, when relative numbers of applications for registration are considered according to "breed."

If particolored varieties are excluded, there are three color types of Cockers recognized by the Standard, and they form the vast majority of the dogs bred, registered, and exhibited in this breed. These are blacks, reds, and blacks with tan-points.

As far as the A series of alleles is concerned, the genes A^s and a^t have been definitely identified. The question is whether an a^y allele also exists. In the 5,569 red puppies recorded by cooperators there is no mention of sable, which indicates that, if a^ya^y or a^ya^t Cockers exist, they may be indistinguishable in appearance from some of the yellow-red types produced by ee action in that breed.[1]

---

[1] Phillips (1938) has described Cockers which are apparently sable a^y with dark hairs in their coats.

It would seem, therefore, that the most promising evidence for the occasional incidence of an animal which was red because of a^y would be an increased percentage of dark-pigmented animals, i.e., black, liver, or bicolor, in red × red matings, as compared with the percentage of these types occurring in breeds where all reds are evidently *ee* in constitution. Such dark-colored animals would be formed in matings of $a^y a^y$ or $a^y a^t$ reds with *ee* reds.

The most extensive data obtained from cooperators bearing on clearly recognized *ee* breeds are those on Irish Setters and Golden Retrievers. If these data are totaled, 1,574 pups were produced by presumably *ee* × *ee* matings. Of these only one (or 0.06 per cent) is of any color other than red-yellow (the *ee* type).

On the other hand, of a total of 3,476 progeny from red × red matings in Cockers, 90 (or 2.6 per cent) are other than red or yellow. It seems probable, therefore, that red Cockers are occasionally produced due to the action of a^y, although commonly their red is of the *ee* type.

This probability is further strengthened by the fact that black tan-pointed animals have been recorded from red × red matings. This would be expected if the parents were $a^y a^t$ in constitution.

Other Cocker data derived from cooperators are given in Table 17. The great majority of red or yellow Cockers seem to be produced by the action of the *ee* combination, and most of these animals are either $A^s A^s ee$, $A^s a^t ee$, or $a^t a^t ee$ in genetic constitution.

Further experiments will be needed to analyze the interaction of *e* and a^y on backgrounds of the CC, Cc^{ch}, and $c^{ch} c^{ch}$ combinations of genes. The appearance of newborn pups recorded as "white" in litters that do not include particolored (piebald) sibs suggests strongly that c^{ch} is present. This may be true in many *ee* breeds.

Cooperators' records of Chow Chow matings are not numerous, but the figures are cited for what they may be worth. In

Table 17. Matings between black *E* and red *e* Cocker Spaniels

Nature of mating	Total pups	Black	Liver	Red	Black-tan-points, bicolor	% red
Black × black	3,520	3,015	7	427	71	12.1
Black × red	5,033	3,128	13	1,787	105	36.2

this breed 22 red × red matings gave a total of 103 pups, of which 100 were red, 2 gray or silver gray (possibly sable), and 1 black and tan. This evidence supports the hypothesis that *aᵁ* as well as *e* reds are produced in this breed.

At the Laboratory a Chow, ♀ 0264, which is a rich, clear red with a black nose, when bred to ♂ 0210, a tan Basenji (*aᵁaᵁEE*), produced a litter of 4 pups, all red. This indicated that ♀ 0264 also was *aᵁaᵁ* or *aᵁaᵗ* in formula. Bred with a black-nosed yellow Dalmatian (*AˢAˢee*), the same Chow bitch in three litters gave a total of 11 pups of some shade of red and *2 solid black*. The appearance of the red pups showed that she must have at least one *e* gene, but the black puppies showed that she must also carry *E* and that she was red because she was *aᵁaᵁ*. This would make her tentative genetic formula *aᵁaᵁEe*. This was tested by mating her to a Doberman (black with tan points; *aᵗaᵗEE*). By him she had a litter of 4 pups, all some shade of the sable-tan-red series, as was expected. Because the total number in the Doberman cross is small, there still remains the possibility that the formula is *aᵁaᵗEe*.

Whitney (1947) gives some interesting data on color inheritance in Chow Chows. Matings of various sorts of blacks or blues between themselves give a total of 123 of the black-blue types and 43 of the red-cream. This is clear evidence of the hypostatic nature of the red-cream genes. On the other hand, matings of animals of the red-cream series *inter se* give 917 of the parental types and 35 with dark pigment (black or blue), an incidence of 3.6 per cent of the latter. These records approach closely the previously quoted figures in Cocker Spaniels (2.6 per cent), where the two kinds of red, $a^y a^y$ and ee, also appear to be present.

A light brindle Greyhound dog of the tentative formula $a^y a^y e^{br} e$ was used at the Jackson Laboratory in a number of crosses. By ♀ 0211, an orange-and-white English Setter ($a^t a^t ee$), he gave, as was expected, 4 brindles ($a^y a^t e^{br} e$) and 4 of different shades of red or yellow ($a^y a^t ee$). By a deep-red Irish Setter, No. 0230 ($A^s A^s ee$), he gave four blacks ($A^s a^y e^{br} e$) and 5 reds ($A^s a^y ee$). *This mating showed clearly that* A^s *conceals the brindle pattern, even if* a^y *and* e *are carried as recessives.* By a liver-and-white Pointer bitch (probably $A^s a^t Ee$) he gave 2 blacks, 2 brindles, and 8 reds, the darkest of which was shaded sable and the lightest cream. Since this cross is more complicated to follow than the two previously listed, it may help to analyze it further (see Table 18). Although in 13 puppies the expectation would be 5 blacks, 5 red-yellows, and 3 brindles, the actual numbers obtained in twelve were 2, 8, and 2, respectively, which may well fall within the limits of chance distribution.

Table 18. Theoretical combinations of genes in E, e^{br}, and e matings
Setter × Greyhound

Gametes of dam	Gametes of sire	Possible progeny	Phenotype
$A^s E$	$a^y e^{br}$	$A^s a^y E e^{br}$	Black
$A^s e$	$a^y e$	$A^s a^y E e$	Black
$a^t E$		$A^s a^y e^{br} e$	Black
$a^t e$		$A^s a^y ee$	Red-yellow
		$a^y a^t E e^{br}$	Brindle
		$a^y a^t E e$	Red-yellow (sable)
		$a^y a^t e^{br} e$	Brindle
		$a^y a^t ee$	Red-yellow

A brindle bitch, No. 633, produced by this mating was tested by two matings. By a black, tan-point Doberman ($a^t a^t EE$) she gave 10 pups, 3 brindles, 2 sables, 3 black with tan-points, *2 black with brindle-points. The blacks with brindle-points, instead of clear tan-points, were a color variety not seen in either parent breed and showed that brindle is not an allele in the A series.* This type has since been obtained in a number of other matings, which will shortly be described. The results obtained (Table 19) show that ♀ 633 was $a^y a^t E e^{br}$ in formula.

Table 19. Theoretical combinations of genes in E, e^{br} and e matings, F₁ (Pointer × Greyhound) × Doberman Pinscher

Gametes of ♀ 633	Gametes of sire	Genotype of progeny	Phenotype of progeny
$a^y E$		$a^y a^t EE$	Sable-tan
$a^y e^{br}$	$a^t E$	$a^y a^t E e^{br}$	Brindle
$a^t E$		$a^t a^t EE$	Black, tan-points
$a^t e^{br}$		$a^t a^t E e^{br}$	Black, brindle-points

Since it has already been demonstrated that animals with one gene for A^s are solid dark-colored individuals, it follows that *brindles are usually, if not always, $a^y a^y$, or $a^y a^t$, in formula.* Brindles which are also $a^t a^t$ have solid dark-colored bodies and show the brindling *only on the points.*

With this in mind, a series of crosses involving one parent which was a brindle Scottish Terrier, No. 0258, was made. This dog, crossed with black-tan-point females ($a^t a^t EE$; Beagles and Dachshunds), gave a total of 10 puppies, all brindle. If we assume that the Scottish Terrier is $a^y a^y e^{br} e^{br}$ in formula, all the resulting F₁ animals are $a^y a^t E e^{br}$. When crossed among themselves, the following types should be produced: 9 brindles, 3 sable-tans, 3 blacks with brindle points, 1 black-and-tan. Actually, the numbers were 11 brindles, 6 sable-tans, 3 blacks with brindle points, and 5 black-and-tans. Although the last-named group is in excess of expectation, *the occurrence of sables* and the absence of intermediate or confusing types support the hypothesis under consideration.

Further supporting evidence is obtained by the backcross of F_1 $a^ya^tEe^{br}$ brindles to a^ta^tEE black-with-tan-point animals. Here the expectation is equal numbers of the four classes. Actually 9 brindles, 5 sables, 2 blacks with brindle points, and 6 blacks with tan points have been obtained. If we remember that in both these crosses there is a possibility that the amount of dark pigment on the points of black-brindle-point animals may be so slight as to make detection difficult, or perhaps impossible, by direct observational methods, the agreement with expectation is not unsatisfactory.

It is interesting to note that, if Scottish Terriers are a^ya^y in formula, they lack the A^s allele in brindle individuals. On the other hand, solid-black individuals are common in this breed. It is uncertain whether some of these are due to a mutation from a^y to A^s, but it would be a matter of interest to know whether strains of Scottish Terriers breeding true to black can be isolated.

Melanotic specimens have been reported in wild Canidae, and it may well be that these represent dominant mutants which are A^sa^y in constitution.

Relation of a^y *to* e. As far as is known, no extensive crosses involving the interrelationship of the a^y and e factors have been reported. In other words, it is not known how compatible the combinations of a^ya^yee may be in a single individual. In a cross between an a^ya^yEE breed such as Basenjis and an A^sA^see breed such as red Cocker Spaniels, the F_2 generation has shown a significant deficiency of red or tan animals. Details on this cross will be published later. It is, however, interesting that in a total of approximately 200 F_2 animals, the ratio of blacks to reds is significantly different from the 9 to 7 expected, if all combinations of a^ya^y and e survived. On the theory that all, or many, animals homozygous for a^y or for ee which possess one or more units of the other gene die, a 9-to-2 ratio would be expected. The observed figures at present approximate this ratio closely.

Should this hypothesis be correct, the majority of red or tan animals obtained in F_2 would be either $a^y a^y EE$ or $A^s A^s ee$, like the grandparental strains. It is hoped that a breeding test of this hypothesis will be possible in the future.

The mutation which produced the so-called wheaten variety in Irish Wolfhounds, Scottish Terriers, and Cairn Terriers may be a change from e^{br} to e. Wheaten would then be $a^y a^y ee$ and would not produce brindles when crossed with $a^y a^y EE$ animals. On the other hand, wheaten might be the result of a mutation from e^{br} to E, in which event wheatens would be $a^y a^y EE$. In Irish Wolfhounds, wheatens crossed together do not produce brindles, according to Darling and Gardner (1933). This suggests that the e mutation is the correct explanation. Further breeding tests to determine which is the correct explanation could and should be made.

Comparison of the A *and* E *series with Other Forms.* The fact that the A series does not follow a steady and progressive line of dark-pigment reduction in dogs, as it does in mice, for example, is of interest.

In mice, the top member of the A series of alleles is A^y, which ordinarily changes all the lower members of the series to clear-yellow coat color. This allele is followed by A^w (white-bellied agouti), which, although ordinarily much more pigmented than A^y, has less dark pigment than the wild-type A, which lies next below it. A, in turn, has less dark and more yellow than black-and-tan (tan-point) a^t. It is interesting to observe how much the tan-point mouse resembles, superficially, the tan-point dog. Below a^t in mice lies a (nonagouti) from which almost all if not all light-pigmented areas have disappeared. The series is orderly and progressive. There is no reported allele of the A series in dogs which corresponds with the a allele of mice.

In dogs, A^s is really a dominant darkening factor, for if E or

e^{br} allows any dark pigment to develop in the hair, A^s commonly spreads that pigment over all or nearly all of the coat and throughout the individual hairs as well.

When the next allele, a^y, is considered, we find that clear-tan animals can be and are produced in the $a^y a^y EE$ forms. Thus there has been a progressive increase in yellow-red pigment in the series of allelic members, which is exactly the opposite of the condition in the *A* series in mice.

Below a^y is a^w (wild color), which allows almost complete dark pigmentation at birth, but this pigment is progressively lost as the individual grows older. In the adult an agoutilike coat of banded hairs, evenly distributed, is formed.

The next allele, a^t (tan points), is an interesting one. It reverses the trend of pigment formation and allows dark pigment to remain on most of the surface of the body. When $a^t a^t$ animals are also *EE* in formula, the dark areas are a clear solid black or liver in color and the light areas are clear tan, red, or cream. When, however, $a^t a^t$ is combined with Ee^{br} or $e^{br} e^{br}$, *the tan, red, or cream areas show brindling, while the dark areas are unaffected by the* e^{br} *gene and remain solidly dark-colored.*

We know that the presence of an A^s allele suppresses brindling, which is due to e^{br}, and one is therefore tempted to assume that in $a^t a^t$ individuals the dark areas allow the type of pigment suppressor characteristic of A^s to be active locally.

Possibly the occurrence of dark pigment in the coats of deeply pigmented sables $(a^y a^t)$ indicates that, although ordinarily hypostatic to a^y, the a^t gene is directly or indirectly responsible for quite a bit of activity in counteracting the restriction of dark pigment, which is the prime function of the a^y gene.[2]

There is an interesting parallel between the interaction of the *A* (agouti) and the *ej* (tortoise shell or brindling) gene in rabbits and the a^t (tan-point) and e^{br} (brindle) genes in dogs.

---

[2] In Dachshunds, animals clear tan from birth have repeatedly been found to carry a^t in breeding tests at the Jackson Laboratory. This is, of course, an exception to the condition described above.

In rabbits, animals that are *aaejej* in constitution are brindled with areas of black and yellow. Those that are *AAejej* might reasonably be expected to have areas of *agouti* and yellow in a brindled distribution. Instead, however, the areas which should be *agouti* are *black*. In rabbits, there is, in addition to E (extension of dark pigment over the whole coat), a dominant black, which has been considered a fourth allele at the e locus and the top member of the allelic series. It would seem probable that this gene is active in the dark areas of the brindle rabbit and that it covers and masks the agouti pattern. In much the same way, the A^s gene, which is able to conceal and cover e^{br} brindling in dogs, is probably operative in the dark area of the $a^t a^t$ tan-point allele. Since the a^y and not the A^s gene is active in the tan areas of this pattern, the e^{br} gene is able to express itself in those areas just as it does over the whole body of $a^y a^y$ or $a^y a^t$ animals.

CHAPTER VII

Locus *G*, Locus *M*, and Locus *P*

THE *G* LOCUS

ALTHOUGH direct experimental evidence on this gene, *G*, is scarce, it would appear that it produces a progressive graying from birth until old age or throughout life. It seems to be dominant over its allele, *g*, which is found in normal dark-pigmented animals, which show no progressive graying.

G expresses itself with considerable variation, both in the time and the extent of its activity. Pups which are *Gg*, and probably at least some of those which are *GG*, are born black, or nearly so. This is seen in Bedlington and Kerry Blue Terriers and in Poodles, both Standard and Miniature.

In Poodles there is and will long remain a great deal of confusion between different types of dilute or pale-colored individuals. This confusion is due to the presence and probable interaction of a number of genes, as well as to a diversified and inconsistent terminology. In the first place, the *B–b* alleles are both present in Poodles, and this means that the chinchilla c^{ch} gene may well have a lightening effect on *bb* or even possibly on *Bb* individuals, which it does *not* have on homozygous blacks, *BB*. The *d* gene for blue dilution is present in some families or strains. This form of dilution, which expresses itself in newborn pups, is different from *G* but might theoretically be expected to supplement or increase the action of the latter if both were present in the same animal.

It is probable, therefore, that controlled genetic studies within the Kerry Blue breed, which is *BBDD* in constitution, and of crosses between this breed and others selected to introduce b and d under planned and purposeful conditions will be needed before a complete and accurate analysis of G can be made.

THE M LOCUS

The M locus contains a pair of alleles. Of these, m, which produces the uniform pigmentation characteristic of most breeds of dogs, is recessive to M, a pattern factor which one finds in merle Collies or Shetland Sheep Dogs, harlequin Great Danes, and dappled Dachshunds or Dunkerhunds (Wriedt, 1925).

The M gene has two major effects. First, it produces irregular blotches of dark (black or liver) pigment against a distinctly lighter background of the same general basic pigment. This it does in either A^s or a^t animals. In $a^y a^y$ (yellow-tan) or in $a^y a^t$ sables the contrast between the dark and light areas of the coat is often discernible only at birth or soon thereafter. As the coat gets lighter with age, the contrast disappears, leaving the $Mma^y a^y$ animal indistinguishable in appearance from the $mma^y a^y$ unless blue sectors in one or both eyes are present. This fact has been recognized by Mitchell (1935) and others in Collies. Work at the Laboratory has shown that it applies to Dachshunds as well. No observations have been recorded regarding any effect of the merle pattern, as contrasting areas of light and dark pigment, in the tan regions of merled tricolored dogs or of merled blacks with tan points devoid of white.

Second, M usually increases somewhat the amount of white spotting on the coat. Although this influence is often observable on self or nonspotted individuals, it is perhaps most noticeable in piebald animals. A double dose of M has far-reaching and usually deleterious effects. In Collies and Shetland Sheep Dogs the MM type is commonly all white, deaf and/or blind, and often sterile as well. In Dachshunds more than half the coat is usually white, and reduction of eye size and function may be present. In Great Danes the situation seems to be much like that in Collies. Table 20 shows the results of the matings ob-

Table 20. Matings involving M and m genes

Nature of mating	No. of litters	Total pups	Observed		Expected	
			Merle	Nonmerle	Merle	Nonmerle
MM × *mm*	4	18	18	0	18	0
Mm × *mm*	24	112	53	59	56	56
Mm × Mm	10	45	33	12	33	11
mm × *mm*	197	1,145	0	1,145	0	1,145
Total	235	1,320				

served at the Laboratory. Segregation and recombination of the *M–m* pair seems to be proceeding in a regular and ordinary Mendelian fashion.

A few data can be added from cooperators' records; these are derived from Collies and Shetland Sheep Dogs. All are from matings of merle (*Mm*) individuals with nonmerle (*mm*). Many of the matings involve a sable nonmerle parent ($a^y a^y mm$ or $a^y a^t mm$). When in combination with $a^t M$ gametes, the sables produced are $a^y a^t Mm$ and, although merle in genetic composition, they may be free from contrasting areas and indistinguishable in appearance from ordinary sables, with which they are included. As a result, the animals classed as merle are usually $a^t a^t Mm$. This means that there should be a departure from the 1-to-1 ratio between merles and nonmerles, because of the excess of apparent nonmerles when the results are tabulated. That such is the case is evident from Table 21. Since, however, the 1-to-1

Table 21. Matings of *Mm* × *mm* animals in two breeds

Breed	No. of matings	No. of pups	Merle	Nonmerle
Shetland Sheep Dogs	5	21	5	16
Collies	9	70	25	45
Total	14	91		

Fig. 5. The merle pattern (M) on $a^t a^t$ animals (tan-point). On the top row are shown smooth-coated Collies, which are heterozygous (left) or homozygous (right) for the M factor. Collies are either s^p (piebald) or s^i (Irish spotting) in formula. They may be contrasted with Dachshunds (lower row), which are SS (homozygous self) in constitution. The left-hand Dachshund is heterozygous merle (Mm). The middle and right-hand Dachshunds are homozygous merle (MM). The latter two figures show the range of pigmentation versus white areas in the homozygous type and may be contrasted with the homozygous (MM) Collies in order to observe the comparative effects of S and its alleles s^i and s^p.

ratio is very closely approximated when the matings observed at the Laboratory are considered, we may conclude that the *Mm* alleles are well established.

THE *P* LOCUS

At this locus we may hypothesize a single pair of genes. The dominant allele, *P*, affects the depth of pigment which characterizes the ordinary color varieties of dogs. The recessive allele, *p*, is considered analogous to a similar gene in rodents. In rodents, the action of *p* radically reduces black and brown pigment, but leaves yellow or tan pigment unaltered. As a result, color varieties of mice in the reduced blacks (*pp* types) have been described as lilac or silver, and in the reduced browns as silver-fawn or silver-champagne.

The occurrence of a parallel variation in dogs has been extremely infrequent. There seems to be fairly clear evidence, however, of ruby-eyed or pink-eyed Pekingese with pale-colored coats of different shades (Pearson and Usher, 1929). It is probable that at least some of these animals are *pp* in genetic constitution. As yet, however, there is no well-established line or strain of this color type from which satisfactory data to determine genetic ratios can be obtained.

CHAPTER VIII

Locus *S* and Locus *T*

THE *S* LOCUS

THIS is one of the most interesting of the coat-color loci in dogs. Affecting as it does the distribution pattern of colored or of white areas on the body surface, it has certain qualities which define and limit its activities. These are similar or analogous to those of the *S* locus in other carnivores or even in rodents. Some of these general characteristics can be advantageously discussed before a more detailed analysis of the situation in dogs is attempted.

Modifying Factors. At each level of white spotting or pigment distribution characteristic of each of the alleles at this locus, various genetic factors independent of the main gene are operative. These cause the extent of pigmented or white areas to fluctuate around a given mean of distribution which the main gene attempts to produce.

Since, in general, an extensively pigmented body surface depends upon alleles that are epistatic to those producing less pigment, we may designate the modifiers producing much pigment as "plus" and those producing or allowing production of less pigment as "minus." The mean amount of pigmented surface can usually be modified by selective genetic breeding in successive generations. This indicates the genetic nature of some of the modifying influences.

On the other hand, an appreciable amount of variation in the extent of body-surface pigmentation is usually nongenetic in nature. This problem has been carefully and skillfully studied and analyzed by Wright in a long series of experiments with inbred strains of guinea pigs and their hybrids. Age, sex, and common uterine environment were among the influences he measured. Even after all correlations with controllable factors had been measured and evaluated, Wright found that unanalyzable and unpredictable individual variation still remained. This he attributed to an "unanalyzable residue" of causative factors, and there we must, at present, leave the matter in dogs as well as rodents.

Limitation of S *Gene's Activity.* The fact that genes which influence spotting produce regions or areas of white in sharp contrast with adjoining pigmented areas is important because it affords an opportunity to estimate and record quantitative variations in the expression of the pigment-forming activity of the various alleles at the S locus. The eye readily recognizes, and mechanical methods can readily detect and record, minute changes in the proportion of colored and white body-surface areas. This is not true of equally small variations in the total amount of pigment, when such pigment is uniformly distributed over the entire body surface or when it has no contrasting pattern in which border lines of different pigments exist.

The importance of this mechanical factor in affecting our description and analysis of gene action has already been emphasized in the discussion of the brindle, e^{br}, allele at the E locus.

It is possible for the action of modifying genes and other influences to be too completely and too sensitively recorded in the response of an animal. In these cases, separation of causative factors and subsequent analyses of them are naturally difficult.

Alleles at the S *Locus in Dogs.* At least four alleles appear to be reasonably well established at the S locus. These, in order of epistasis, are as follows:

(1) S —self, or completely pigmented body surface;
(2) s^i —Irish spotting, with few and definitely located white areas;
(3) s^p —piebald spotting;
(4) s^w —extreme-white piebald.

We shall now discuss these alleles and give some consideration to the modifiers which influence their expression. Diagrammatic illustrations of the action of plus and minus modifiers are shown for S animals in Scottish Terriers, for s^i in Basenjis, for s^p in Beagles, and for s^w in Sealyham Terriers.

s^w—*Extreme-white Piebald*. This is the lowest member of the series of four alleles at the S locus. In its clearest and least confused expression it is seen in such breeds as Bull Terriers, Samoyeds, Great Pyrenees, Sealyham Terriers, Greyhounds, and Bulldogs.[1]

Fig. 6. Extreme piebald (s^w) is found in a breed like the Sealyham. Those with minus modifiers are pure white, while those with plus modifiers have small pigmented areas and overlap with piebalds (s^p) having minus modifiers.

Cooperators' records available for study at the Jackson Laboratory cover litters totaling in all 77 Bull Terrier pups, 110 Samoyeds, 29 Great Pyrenees, and 43 Sealyham Terriers. The distribution of individuals of these groups according to coat pigmentation is shown in Table 22.

---

[1] It is also, in all probability, characteristic of Dalmatians, but since the gene T for ticking is present in that breed and produces one of the most characteristic features of the Dalmatian's pattern, the consideration of the contribution of s^w to that pattern will be postponed until the T locus is discussed.

Table 22. Distribution of pigment in extreme piebalds (s^w) of several breeds

Breed	Total pups	All white	1–9% colored	10–19% colored	% all white	% 1–9% colored	% 10–19% colored
Samoyed	110	101	9	0	90.9	9.1	0.0
Bull Terrier	77	63	14	0	81.8	18.2	0.0
Great Pyrenees	29	4	18	7	24.1	62.1	13.8
Sealyham Terrier	43	17	26	0	39.5	60.5	0.0
Total	260						

In two other breeds where all-white or extreme-white piebald animals are found frequently with no apparent relationship to the *M* (dapple) gene, there is evidence that s^w is present. These breeds are Collies and Boxers. Because a relationship between s^w and one or more other alleles at the *S* locus is involved in each of these breeds, they will be further considered under the Irish and self alleles, respectively.

In Collies and Boxers the records are sufficient to provide satisfactory evidence of segregation of s^w from other *S*-locus alleles. In Greyhounds and Bulldogs the existence of all-white or very nearly white individuals is sufficiently frequent to make it highly probable that s^w is present in both cases.

Warren, in an article on the genetics of coat color in Greyhounds (1927), stated that more extended white spotting is recessive to heavier pigmentation and that "white dogs are an extreme of spotting." This is consistent with the behavior of s^w in other breeds where its presence has been established.

s^p—*Piebald Spotting.* The amount of spotting that can be produced in animals with the genetic formula $s^p s^p$ is very variable. The typical expression of this allele in the *S* series is seen in a breed such as the Beagles. Fig. 7 shows a series of grades of

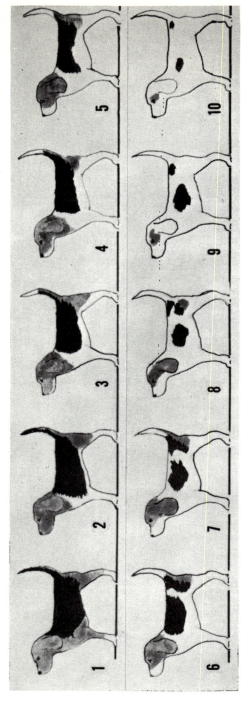

Fig. 7. Beagles are used as an example of the extreme range of the piebald pattern, which is determined by the s^p gene. A series of arbitrary grades (1–10) has been given to the different degrees of pigmented versus white areas. The heavily pigmented grades are considered to be due to "plus" modifying genes acting on s^p. The relatively unpigmented types are due to "minus" modifiers. Cf. Graph 1. This figure should be compared with Figs. 6–9. It will be noted that Grade 1 overlaps with SS animals having minus modifiers (Fig. 9); Grades 1–4 overlap with the Irish s^i pattern (Fig. 8). Grade 10 overlaps with the plus-modifier extreme-piebald pattern, s^w (Fig. 6).

increasing proportion of white areas arbitrarily picked and numbered from 1 to 10. A distribution of 816 Beagle pups listed in cooperators' records according to the percentage falling at, or near to, each of these grades is given in Table 23 and Graph 1. This forms a skew curve with a remarkably steady decrease in the number of pups as more and more white occurs.

Table 23. Distribution of pigment in piebald (s^p) Beagles

Grade	Percentage
1 (least white)	55.9
2	19.6
3	8.3
4	5.4
5	3.2
6	2.8
7	1.4
8	1.3
9	1.4
10 (most white)	0.7

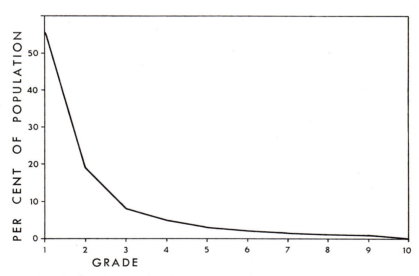

Graph 1. The distribution of grades in a population of 816 Beagle pups. Cf. Fig. 7.

In experiments at the Jackson Laboratory 75 litters have been produced from piebald ($s^p s^p$) × piebald ($s^p s^p$) matings. All of the 428 progeny have been piebald. Only one mating of piebald carrying extreme-white piebald ($s^p s^w$) × extreme-white piebald ($s^w s^w$) has been made. Since piebald is epistatic to extreme-white piebald, a 1-to-1 ratio was expected. Actually, 5 piebald and 7 extreme-white piebald animals were obtained. In a distribution such as that obtained with the Beagle puppies, it is probable that both modifying genes and nongenetic influences affect the amount of pigment formed.

Data on other piebald breeds from the cooperators' blanks show how this pattern reproduces itself constantly (see Table 24).

Table 24. Results of s^p × s^p matings in several breeds

Breed	Total	Piebald	Irish	Self
Brittany Spaniels	85	85	0	0
Wirehaired Fox Terriers	90	90	0	0
English Setters	675	675	0	0
Cocker Spaniels	537	527	7*	3*
Springer Spaniels	344	344	0	0
Total	1,731	1,721	7*	3*

* It is probable that although the parents of these animals were classed as "particolor" (piebald) they were actually self animals, either Ss^p or Ss^w in formula. As will be more fully discussed under the S allele, the dominance of that gene is often incomplete in Cockers, and pseudo-piebalds may thus be formed. It is important to note that this is purely a phenotypical or synthetic appearance of the Irish pattern, similar in somatic type but of a very different genetic constitution from that of $s^i s^i$ animals.

s^i—*Irish Spotting.* One of the earliest descriptions of this pattern was given by Doncaster (1905), who found that it appeared characteristically in Norway rats, which were heterozygotes between S, the gene for solid or self coat, and s^h, the gene for the "hooded" or piebald pattern.

The pattern consists of white spots or streaks in one or more of the following locations: (1) muzzle, (2) forehead star or blaze, (3) chest, (4) belly, (5) one or more feet, (6) tail tip. In development under selective breeding, spots which may fuse and extend to produce a collar can also be produced. Certain of the more restricted degrees of the pattern were frequently found in wild rats from Ireland, which, in all probability, caused Doncaster to label this spotting "Irish."

The first clear demonstration of s^i as a gene allelomorphic to S and to s^p was provided by Castle and Phillips (1914). They had been studying variation in the $s^h s^h$ genotype of Norway rats under selection. The Irish pattern in its typical form appeared in a single individual as a mutation from s^h to s^i. Since s^i is epistatic to s^h, the mutant was $s^i s^h$ in formula. When crossed with $s^h s^h$ animals, the progeny consisted of Irish and hooded animals in two clearly distinct groups with no confusing or intermediate grades of spotting. Further descendants of the cross gave ratios of the two types with clear-cut segregation characteristic of a monohybrid relationship between them.

Consideration will be given later to the close parallelism that exists between rats and dogs in respect to the creation of pseudo-Irish spotting in heterozygotes between S and s^p. For the present, however, attention will be confined to the s^i gene as such and to its behavior in crosses.

One of the most typical and consistent examples of the Irish pattern in dogs is found in Basenjis. The extent of the pattern as affected by genetic modifiers and nongenetic influences is shown for this breed in Fig. 8. An interesting mutation from S to s^i in a German Shepherd puppy has been reported to the writer by Mr. Clarence Pfaffenberger. The German Shepherd breed usually has only extremely small white spots or is solid-colored with no white at all. The distribution of white on the mutant just mentioned is roughly the same as in the most heavily pigmented type of Collie in which the Irish pattern is developed. The pattern is epistatic to both s^p and s^w but is hypostatic to S, the gene for self or complete pigmentation.

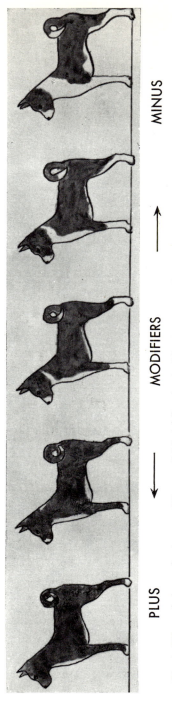

PLUS ← MODIFIERS → MINUS

Fig. 8. Effect of plus and minus modifiers on Irish ($s^i s^i$) animals. The Basenji is of that genetic formula. Note that piebald ($s^p s^p$) animals overlap with this series (Fig. 7).

The results of crosses involving s^i and its hypostatic allele, s^p, which have been observed at the Jackson Laboratory are given in Table 25. There is evidence that the usual type of

Table 25. Matings involving s^i and s^p genes

Nature of mating	No. of litters	No. of pups	No. Irish	No. piebald
$s^i s^i \times s^i s^i$	24	145	145	0
$s^i s^i \times s^p s^p$	18	101	101	0
$s^i s^i \times s^i s^p$	6	36	36	0
$s^i s^p \times s^i s^p$	15	93	71	22
$s^i s^p \times s^p s^p$	5	29	17	12
Total	68	404		

Collie, with white feet and a white collar of varying width, may represent the Irish type of spotting. The relatively consistent location of the white areas suggests one of the characteristic features of Irish Spotting in contrast to the irregular extent and distribution of colored and white areas seen in piebald animals such as Beagles, Greyhounds, and Springer and Cocker Spaniels.

A similar situation probably exists in Boston Terriers, where the vast majority of animals have white areas in essentially the same locations as do Basenjis. In Boston Terriers and in Collies the s^w allele also occurs. It is undesirable in the former and has been rigorously selected against.

The cooperators' blanks report a total of 364 Boston Terrier puppies showing "Irish" spotting, and 7 are recorded as "almost all white" or "white." There are no records of animals with the intermediate or irregular spotting characteristic of most piebalds.

In both Boston Terriers and Collies there is a strong tendency for increased amounts of white in the piebald grades between the Irish and extreme-white piebald patterns to follow a regular distribution. This is an extension of unpigmented areas in the location of those areas in the Irish pattern. Their distribution does not appear to be as irregular and unpredictable as they are in piebald s^p animals such as Brittany Spaniels, Springer Spaniels, or Pointers. This suggests that Boston Terriers and Collies of the relatively rare intermediate degrees of white spotting may be due ordinarily to incomplete dominance of s^i over s^w under the influence of minus modifiers.

In Collies there is an interesting distribution of the amount of white spotting in a total of 215 animals recorded by the cooperators. In this tabulation the same arbitrary grades described for Beagles were used. The percentages of the population falling in each grade are shown in Table 26. There are

Table 26. Distribution of pigment in piebald (s^p) Collies

Grade	Per cent
1 (heavily pigmented)	0.0
2	0.9
3	37.67
4	10.70
5	4.19
6	6.05
7	6.92
8	11.16
9	19.07
10 (white or with only traces of pigment)	3.26

clearly two modes, as underlined. Grade 3 may be classed as typically Irish spotting, grades 8, 9, and 10 as extreme-white piebald. The small numbers in grades 5, 6, and 7, the intermediate types, and the regularity of the spotting in the animals of those grades suggest, as above mentioned, that they may well be $s^i s^w$ heterozygotes with minus modifiers and with incomplete dominance of s^i over s^w, rather than genetically piebald, s^p.

In many breeds animals occur showing essentially the same amount and the same distribution of white spotting as those typical of the Irish pattern. Among these breeds may be listed Cocker Spaniels, Gordon Setters, and others in which such animals are considered as being "mismarked" from the point of view of the bench show. Study of the conditions under which such mismarked animals occur leads to the conclusion that they are not $s^i s^i$ in constitution except in very rare instances. They appear to be animals with at least one self, or S, gene present. They will, therefore, be considered under that heading, which follows.

S—*Self Color.* In breeds that are usually SS in constitution, "no white" is to be considered the typical condition of coat pigmentation. However, just as Irish spotting ($s^i s^i$), piebald spotting ($s^p s^p$), and extreme-white piebald ($s^w s^w$) show the effect of plus or minus modifiers, so the SS breeds in the presence of minus modifiers may develop a certain amount of white. This is shown diagrammatically for Scottish Terriers in Fig. 9.

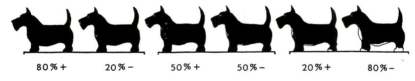

| 80% + | 20% − | 50% + | 50% − | 20% + | 80% − |

Fig. 9. Effect of plus and minus modifiers on SS (self) animals. The Scottish Terrier is almost universally of this genetic type. Note that the SS animals with minus modifiers overlap with the Irish animals with plus modifiers (Fig. 8).

The similarity of the effect of the extreme minus modifiers of SS and the phenotypical appearance of Irish spotting ($s^i s^i$) is at once obvious.

A tabulation of cooperators' records giving data on the incidence of mismarked or pseudo-Irish individuals in a number of self (SS) breeds is interesting (see Table 27). It may be taken as an indication that breeds differ in the proportion of plus and minus modifiers possessed and that the influence of these modi-

Table 27. Incidence of "mismarked" (peudo s^i) animals in various breeds

Breed	Total pups	Self	Mismarked or pseudo-Irish	% self
Schipperke	69	66	3	95.6
Dachshund	661	622	39	94.1
Scottish Terrier	520	467	53	89.8
Labrador Retriever	734	649	85	87.0
Irish Setter	249	201	48	80.7
Pug	77	61	16	79.2
Newfoundland	43	21	22	48.8
Total	2,353	2,087	266	88.7

fiers is still felt, even in breeds where the effect of minus modifiers is considered to be undesirable. This, in turn, may mean that because of the number of such modifiers, and/or their failure to express themselves in an easily recognizable and consistent manner, their elimination by selective breeding is and will continue to be a difficult matter.

Data derived from actual matings at Bar Harbor, involving action of the gene S, are shown in Table 28. It is evident from

Table 28. Matings involving S, s^i, s^p, and s^w genes

Nature of mating	No. of litters	No. of pups	S	s^i	s^p	s^w
$SS \times SS$	54	262	260	2	0	0
$SS \times Ss^p$	3	27	15	12	0	0
$SS \times s^i s^i$	3	14	14	0	0	0
$SS \times s^p s^p$	29	180	46	134	0	0
$Ss^p \times Ss^p$	2	10	1	8	1	0
$Ss^p \times s^p s^p$	4	31	1	13	17	0
$Ss^w \times s^p s^w$	1	8	0	4	3	1
Total	96	532				

these figures that the heterozygous Ss^p type is usually mismarked with a "pseudo-Irish" degree of spotting that represents incomplete dominance of S over the s^p allele. The extent to which dominance is complete is probably determined by the kind of modifiers of the S-locus genes involved.

Two popular and widely distributed breeds in which segregation of the S gene from its other alleles may be observed and studied are Boxers and Cocker Spaniels. In Boxers self or solid-colored animals, those with an Irish or psuedo-Irish degree of spotting, and those which are all white or which have a very small amount of pigmented area are recorded. The distribution according to white areas occurring in 1,780 Boxer pups listed on cooperators' blanks is an interesting one. It is shown in Table 29. The bimodal nature of the curve is evident. The dis-

Table 29. Distribution of pigment in Boxers

Amount of coat pigmented	No. of pups	% in each class
100% (self)	330	18.30
90–99% (Irish)	1,313	73.70
80–89% ''	24	1.35
70–79% ''	0	0.00
60–69% ''	1	0.06
50–59% ''	1	0.06
40–49% ''	1	0.06
30–39% ''	3	0.18
20–29% ''	2	0.12
10–19% ''	3	0.18
0–9% ''	102	5.73
Total	1,780	

tribution of spotting grades clearly suggests the presence of three alleles in the S series, namely, S (solid coat color); s^i (Irish spotting); and s^w (extreme-white piebald). It is probable that the exceptional animals appearing to have intermediate grades are due to the action of modifiers on s^i in a minus direction or on s^w in a plus direction.

The Inheritance of Coat Color

In Cocker Spaniels a very interesting and somewhat complex situation occurs. Both solid-colored and spotted (parti-color) varieties are admitted and encouraged by the Standard for this breed. But animals that have the Irish type of spotting are considered mismarked and are rigidly discouraged. In spite of this fact, they occur in large numbers in the breed and are seeded throughout all or nearly all blood lines being produced by fanciers. They are produced by two, and possibly by three, genetic processes: (1) They may be due to modifying genes acting on SS (solid-colored) animals. (2) They may be due to incomplete dominance of S over s^p or s^w in Ss^p or Ss^w individuals. Both of these conditions are clearly and unmistakably encountered. A third possibility is that the s^i (Irish-spotting) allele may also be present. The likelihood of this, however, is small, for it would be selected against rigidly if it occurred in SS lines and could only persist for any length of time if it coexisted in Ss^p or Ss^w families. Here it might well be indistinguishable in appearance from those types and so be carried along without the breeder being aware of its presence.

A tabulation of the different amounts of pigment on the coat has been made (Table 30). Graph 2 shows the curves of percentage distribution of 385 black and 432 red pups (total 817). The solid line shows that blacks occur in gradually decreasing frequency from the heavily pigmented animals at one end of the curve to the lightly pigmented ones, of which only a few are recorded. The distribution of spotting in red animals, on the other hand, shows that the great majority (80–93 per cent) are grouped in the classes having 30 per cent or less pigment. Whitney and Burns also report more white in reds than in blacks.

It is probable, therefore, that we are dealing with two alleles of the S locus, namely, s^p and s^w. It also seems probable that there may be a linkage between e, which produces the common type of reds and yellows in Cockers, and s^w, the gene for extreme-white piebald. Whether the heavily pigmented red piebalds and

Table 30. Percentage of colored coat in black and in red Cocker Spaniels according to the type of mating from which the pups were derived

Nature of mating	90-99%	80-89%	70-79%	60-69%	50-59%	40-49%	30-39%	20-29%	10-19%	9-0%	Total
					Blacks						
Black × black	32	28	19	15	15	10	9	5	4	2	139
Black × red	20	20	20	15	12	6	9	4	3	4	113
Red × black	26	17	15	15	13	10	10	7	8	3	124
Red × red	3	..	..	..	1	1	1	..	1	2	9
Composite	81	65	54	45	41	27	29	16	16	11	385
	(21.04)	(16.88)	(14.00)	(11.68)	(10.65)	(7.01)	(7.54)	(4.15)	(4.15)	(2.86)	
					Reds						
Black × black	0	1	3	1	2	1	0	5	6	9	28
Black × red	1	3	5	4	6	8	7	7	19	19	79
	(1.26)	(3.79)	(6.32)	(4.95)	(7.58)	(10.11)	(8.85)	(8.85)	(24.01)	(24.01)	
Red × black	2	1	2	1	3	4	11	12	19	17	72
	(2.77)	(1.38)	(2.77)	(1.38)	(4.16)	(5.55)	(15.26)	(16.65)	(26.37)	(23.59)	
Red × red	5	5	2	5	10	7	16	32	71	100	253
Composite	8	10	12	11	21	20	34	56	115	145	432
	(1.85)	(2.31)	(2.77)	(2.54)	(4.85)	(4.62)	(7.87)	(12.9)	(26.6)	(33.56)	

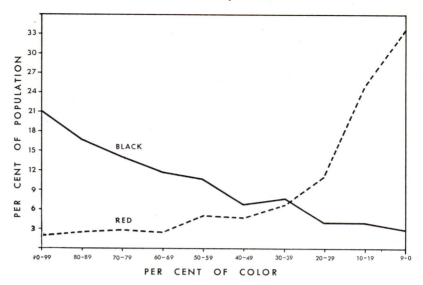

Graph 2. Percentage distribution of pigment on coats of 385 black and 432 red puppies.

the very white piebalds with black pigment are "crossovers" between s^p and E and between s^w and e is not certain. They may, in part at least, be the result of extreme-dark and extreme-light modifiers of piebald (s^p).

If the reader will refer to Fig. 7 (p. 78), in which the various degrees of piebald spotting in Beagles are shown in ten grades, progressing from grade 1, the most heavily pigmented, to grade 10, covering animals with the smallest amount of pigment, he can visualize ten similar grades of spotting in Cockers.

A tabulation of the mean grade of spotting in black and in red Cocker pups in four different categories of matings appears in Table 31. It is interesting to note that both the black and the red progeny of *black* mothers have more pigmentation on the average than do the progeny of *red* mothers. This fact, taken in conjunction with the discussion which follows on the localization of spots in mismarked animals, may indicate a definite metabolic difference between blacks and reds in this breed.

Table 31. Relative amount of white spotting in progeny of matings
of black and red Cockers

Nature of mating	Total black pups	Avg. grade of black pups	Total red pups	Avg. grade of red pups
Black ♀ × black ♂	139	3.59	28	7.67*
Black ♀ × red ♂	113	3.95	79	7.36
Red ♀ × black ♂	124	4.18	72	7.87
Red ♀ × red ♂	9*	5.55	253	8.46

* The number of individuals in this category is perhaps too small to provide reliable data for definite quantitative comparison.

A study of the mismarked animals is interesting. It will be recalled that in many cases the pattern of white spotting closely resembles that of Irish spotting.

Puppies reported by cooperators from matings of one black and one red parent were tabulated according to color on two counts: one of these was the occurrence of mismarking at any location; the other was the incidence of mismarking in the form of a star or spot on the forehead (see Table 32). The smaller

Table 32. Percentage and types of "mismarking" in red and black Cockers

Color of pups	Total pups	Solid, self	Total mismarked	% of Total mismarked	No. of forehead spots	% of mismarked showing forehead spots
Black	1,842	1,228	614	33.3	2	0.32
Red	1,058	793	265	25.1	127	47.90

percentage of total mismarking among reds may not be real. Many of the cream or light-yellow Cockers have so little pigment that the contrast between the ground color and a small amount of white mismarking is so slight that failure to observe the white spot frequently occurs. This is especially true if the white is on the feet and legs where hair is short at birth (when most recordings are made).

Nevertheless there is an extraordinary excess of forehead-spot mismarking among reds as compared with blacks. Whether this is due to a difference in time of activity in the formation of the two colors on the part of gene S during embryonic development or whether it is a form of mismarking which characterizes a dominance relationship between S and s^w is not certain. In either case it probably has a morphogenetic basis. An analysis of the nature of the developmental factors that contribute to this interesting difference in spot distribution between the two color varieties would be of real scientific importance.

This evidence of a morphogenetic difference between blacks and reds may also bear on the greater amount of white commonly seen in reds as compared with blacks.

THE T LOCUS

Ticking with pigmented flecks, or small spots, on a white background is found in a number of breeds of dogs, notably Hounds, Pointers, Spaniels, Setters, and Dalmatians. It may be distributed on any *white* area of the head, body, or extremities, or it may be distinctly more prevalent in those localities where tan pigment occurs in the tan-point pattern. Such a limited distribution, however, is probably not a product of the tan-point pattern itself but may represent an allele of the gene that allows ticking to appear in any white area without restriction as to site.

Ticking confined to the tan-point areas has been observed in animals which are not themselves tan pointed ($a^t a^t$). These individuals have not been tested to determine whether they are $A^s a^t$ in constitution. There remains the possibility that the a^t pattern can influence the appearance of ticking even though present with an epistatic allele A^s in the $A^s a^t$ relationship. There is also the possibility that a morphogenetic relationship exists between A^s, S, and T to the extent that failure of dark pigment to cover the whole surface area, as it does under the action of a^t, is seen in very much the same areas as those which are unpigmented when S fails to cover the whole surface as in Irish spotting (s^i). Selective distribution of ticking usually occurs in these same areas.

One of the general theories to account for patterns which utilize these "extremity" areas is that they represent relatively late points of embryonic development. One gets the general impression that the genes A^s and S are unable to use these areas of late development if and when their own activity is impaired either in amount or by delay.

This cannot, however, be due to *delay in action* of A^s or S, for were such the case the late-developing regions would *receive* pigment instead of *lacking* it. It would seem that where these areas showed tan pigment, as in $a^t a^t$ animals, the A^s gene had completed all of its action early and was not able to function in the areas of late development. The same line of reasoning would apply to the early action of the S gene when it fails to cover the late-developing regions.

In this connection it is interesting to note that the effects of gene T become evident after the birth of the puppy. Dalmatians are born clear white or with a few clear-pigment blotches due to the s^w pattern, as far as coat color is concerned. There is some evidence that ticking, at least on the extremities of breeds other than Dalmatians, may appear earlier than it does in the Dalmatian breed. This may be due to modifiers or some metabolic factors characteristic of the breed.

Some breeders and geneticists are inclined to look upon Dalmatian spotting as being due to a genetic factor distinct from that which produces ticking in ordinary piebald types. Preliminary experiments at the Jackson Laboratory indicate that instead of being piebald, Dalmatians are $s^w s^w$, or extreme-white piebald. This theory is indirectly supported by the fact that the total areas occupied by the larger pigment spots on Dalmatians usually fall within the ordinary range of amount of pigmentation shown by $s^w s^w$ dogs without the ticking factor T.

Matings of the Dalmatian pattern with self-colored dogs recorded at the Laboratory show that ticking occurs in the very restricted areas of white present in the F_1 Ss^w hybrids. The partial dominance of gene S over s^w allows some white areas to appear. In these regions the gene T, which is dominant, expresses itself.

Crosses at this Laboratory between a Dalmatian and a Basenji (a breed in which we have never observed ticking) gave 5 pups all of which showed ticking. The spots appeared at essentially the same time (3 to 4 weeks) that they do in breeds with the T gene, such as Cocker or Springer Spaniels. They were also of the same size and shape as is produced by the T factor in these breeds. This indicates clearly that the Dalmatian introduced the T gene.

In numerous later crosses between Cocker Spaniels known to have the T gene and Basenjis, the F_1 hybrids were of essentially the same type as those produced from matings in which the T gene was introduced by a Dalmatian. It is apparently the same gene in that breed and in other breeds such as Spaniels, Setters, and Pointers.

Dalmatians appear to be homozygous for T, since the white spots on all $Ss^w Tt$ hybrids so far observed have been ticked.

In the F_2 animals there is evidence that S–s^w and T–t segregate independently. Four classes of F_2 animals should be obtained as follows:

Expected ratio

9	$Ss^w TT$ or $Ss^w Tt$	—few white areas, all ticked
3	$Ss^w tt$	—few white areas, non ticked
3	$s^w s^w TT$ or $s^w s^w Tt$	—Dalmatian type, ticked
1	$s^w s^w tt$	—white, or nearly white, not ticked

A cross between an Irish Setter ($SStt$) and a Dalmatian ($s^w s^w TT$) has been carried through the F_2 generation. In the four classes listed the numbers obtained were 31: 12: 12: 4, respectively. The expectation is 33: 11: 11: 4. These data support strongly the explanation proposed for Dalmatian spotting.

Many pattern factors, such as Irish spotting, brindle, or merle, are so slightly expressed phenotypically on individuals as to be extremely difficult or even impossible to identify visually. It may well be that the instances reported of two piebald nonticked parents giving rise to ticked pups will prove to be examples of matings in which one parent, at least, was genetically *Tt* while appearing somatically to be nonticked.

POSSIBLE *R* LOCUS—ROAN

Roan in dogs is a color formed by a mixture of colored and white hairs. In this respect it resembles the type of coat color known as "silvering" in rodents. It exhibits a great deal of individual variability in its expression in dogs as it does in rodents.

The general feeling is that in most cases roan is dominant to clear piebald spotting, but while this may be true commonly, there are enough exceptions and complications to require qualification of such a simple classification of its genetic behavior. Cases have been recorded in Cocker Spaniels in which a puppy born *solid black* has progressively shown an increasing number of white hairs scattered throughout its coat until it finally was very clearly classifiable as a roan. This, however, may be a type of genetic silvering entirely distinct from the roan that occurs on a white area. Similarly, Cockers, born a black parti-color with clear-white areas, have, with increasing age, developed more and more black hairs scattered through the white areas until a definitely roan coat was formed. This procedure resembles the course of development shown by ticking. However, in at least one case the end result was to all appearances a solid-black dog, the white hairs being discernible only after close examination of the undercoat.

There is a question as to whether some relationship or interaction exists between roan and ticking. In such a breed as the

German Short-haired Pointer, it is often extremely difficult to distinguish between what may be a small tick spot and what may be a local condensation of pigmented hair in a roan area.

In English Pointers roan is not common while ticking is much more frequent. In one animal personally observed by the writer, the ticking all over the body was unusually heavy, with a very large number of small, irregularly distributed ticks. On each front leg, located much like a wristlet, was an area perhaps six inches in length in which the pigmentation was clearly black-and-white roan. In view of the rarity of roan in this breed and in the presence of the extremely heavy degree of ticking in this animal, one wonders whether the simultaneous presence of the two conditions has any significance genetically or whether it is merely coincidence.

Because of the lack of definitive data and the clear evidence of genetic complexity in the case of roan, it is probably wise to adopt a conservative but open-minded attitude toward the acceptance of a definitely dominant roan gene at a single locus. Present information makes it seem more likely that ticking and roan grade into one another. They fail to give sufficiently clear-cut segregation to justify separation genetically without additional controlled experimental test matings.

CHAPTER IX

Possible Somatic Mutation

IN SOME of the red breeds produced by the A^*A^*ee combination of genes, instances of what appears to be a somatic mutation of e to E have been reported. A somatic mutation from e to E would produce a localized area of black hairs on a red background. This incidence of an isolated black spot has been observed rather frequently.

A consideration of this phenomenon in Cocker Spaniels will show what the situation is. It will be recalled that breeders' records of red × red parents gave a total of 3,476 pups, of which 3,386 were red (see Table 33). Certain of the animals recorded in the table are probably not true somatic mutations.

Table 33. Record of possible somatic mutations in Cockers (breeders' records)

No. of mating	Total pups	Red	Black	Red with black spots	Location of spots	Other comments
\multicolumn{7}{c}{Nine matings of red × red, producing 9 red dogs with black spots}
CS283	6	5	0	1	Right side	Spot was *white* at birth
CS7924	5	4	0	1	Left side of neck	None
CS7954	7	6	0	1	Right thigh	Very small spot
CS8014	4	3	0	1	Side of muzzle	Very small spot
CS7820	8	7	0	1	Second joint of right rear leg	
CS7812	5	4	0	1	On back near base of tail	
CS7645	4	3	0	1	Right shoulder	
CS7642	4	3	0	1	Two near base of tail on right side	
CS7643	4	3	0	1	Back of left ear	Size of half-dollar

Table 33 (continued)

No. of mating	Total pups	Red	Black	Red with black spots	Location of spots	Other comments
Six matings of red × black, producing 7 red dogs with black spots
CS7107	4	2	1	1	On face	
CS7106	6	4	1	1	Black ear and shoulder	
CS6854	5	1	2	2	On back in both cases	Present at birth, but disappeared later
CS6794	4	2	1	1	Whole left ear and large shoulder patch	
CS7167	8	3	4	1	Left ear all black	
CS7165	6	2	3	1	On right side, near dorsal line	
Four matings of black × black, producing 5 red dogs with black spots
CS6169	8	0	6	2	Around end of mouth, nose, and lips, in both	
CS6459	4	2	1	1	Few black hairs on underside of right ear	
CS6553	11	2	8	1	Small dot on head	
CS6617	6	1	4	1	Ears and near lips	

One of the puppies resulting from mating CS283 had a spot which was white at birth and which later became black. This is the only black-spotted animal on which such a spot is recorded, and the genesis and history of the spot suggests that something more is involved in this case than a direct somatic mutation from e to E.

Two cases of mutation of red to black on the coat have occurred in purebred Cockers at this Laboratory (see Table 34). Note that ♀ 1651 was the daughter of ♀ 1333.

A similar situation has occurred in a cross between a fawn F_1 hybrid (Pointer × Greyhound) and a purebred yellow Dalmatian. There were 7 pups. Six were various shades of red, and

Table 34. Record of possible somatic mutations in Cockers (laboratory records)

Kennel No. of animal showing mutation	Parents	Total pups in litter	Color of pups			Location of spot	Size of spot
			Red	Black	Red with black spot		
♀ 1333	0414 blk. and white × 701 blk.	6	0	5	1	Right in front of ear	Small
♀ 1651	1333 × 0840 red	7	6	0	1	Left shoulder	Small

one (♀ 1390) was red with a small black spot on the left side near the shoulder.

The animals in mating CS6854 had black spots at birth but these disappeared later. If gene *e* mutated to *E*, it would be difficult to explain this result. On the other hand, if these animals had been reds of the $a^y a^y EE$ or $a^y a^t EE$ type, which, although rare in Cockers, may well exist, the somatic mutations might have been from a^y or a^t to A^s. In this case the fact that there is a tendency for a^y animals to become clearer red with increasing age might have a bearing on the disappearance of the spots. One would expect, however, that if this was the correct explanation, substances from the gene a^y that were active in the tissues surrounding the spots must have invaded, or otherwise influenced, the hair follicles underlying the spots. They would then take on the type of pigment-forming activity characteristic of the rest of the animal.

It is interesting to note that mutant red animals from the red × black or black × black matings included a larger proportion of individuals with extensive black spots than did the population derived from red × red matings. Thus in black × black matings, 3 of the 5 animals with spots showed considerable areas of black around the mouth and ears. In black × red matings, out of 7 animals, 3 had relatively large black areas. In the 9 mutant progeny of red × red matings, none were so extensively affected.

In cooperators' records of litters of Irish Setters totaling 943 red pups, one had a small black spot on its side. This is an incidence of 0.106 per cent. In the red × red Cocker matings on cooperators' blanks the incidence is 0.233 per cent. There is probably no significant difference between breeds in the rate of somatic mutation from e to E, for it is so infrequent that enormous numbers of progeny would be needed to analyze the meaning of such quantitative variation in incidence of mutation as may possibly occur. Yet if the a^y gene occurs in Cockers and does not occur in Irish Setters (which seems certain), there would be a chance for *two* genes to produce black spots on reds by mutation in the former breed and for only *one* to mutate in the latter.

Further study of these unusual types of animals would be valuable, as would be microscopic examination of hair from both the black and red areas.

PART THREE

Genetic Analysis by Breeds

Introduction

IN THE following brief reviews of some of the genetic aspects of coat color in various breeds, the system of groups employed by the American Kennel Club has been followed (*The Complete Dog Book*, 1951). Not all breeds are included, because in some the complexities of coat color are so great and the reliable breeding data so meager that postponement of analysis is necessary.

The reader is advised to refer frequently to Part Two, where genes are discussed separately or in relation with one another. Only by such reading and study can a nonscientist become familiar with genetic principles. In many of the sections in Part Two data on specific breeds, as reported by cooperators or as observed at the Jackson Laboratory, are included. References to these data will be found in the index.

The reader is urged to study the discussion of breeds other than those in which he is particularly interested. In this way he can broaden the base of his genetic knowledge and learn to understand the relative position of his own breed or breeds.

It is hoped that breeders with new or interesting data on the incidence or inheritance of mutations, unusual in their breeds, will communicate these data to the Jackson Laboratory. By such cooperation a reasonably complete and up-to-date record of such instances can be assembled and maintained. The occurrence of a mutation producing colors not sanctioned by the Standard for the breed is no disgrace and no evidence of impure ancestry in the animals producing it. It is, on the contrary, an event of real scientific interest and importance.

CHAPTER X

Sporting Dogs

GRIFFONS (WIRE–HAIRED POINTING)

ALL animals of this breed are bb (brown; called chestnut in the Standard). They also seem to be $s^p s^p$ (piebald), varying in their degree of spotting but not ordinarily of either the s^i (Irish) or s^w (extreme-white piebald) type. Their hair length and texture sometimes make it difficult to determine whether ticking (T) is, or is not, present in adults. Probably this gene is present in almost all individuals and only rarely absent.

New born pups of this breed have a clear-white ground color and, if observed continuously, will allow the recording of the appearance of ticking as they increase in age. Col. T. DeF. Rogers, secretary of the Specialty Club of this breed, informs me, in personal correspondence, that in his long experience he has seen only one individual in which ticking had not developed at six months of age.

The genetic nature of the gray, gray-white, or dirty-white background on which the chestnut patches occur and which are mentioned in the Standard is not understood and needs further study. It may well be that the background on which the distinct spots of brown (chestnut) appear is the result of the combined action of a gene for roan which Whitney and others have mentioned and of the gene (T) for ticking. The roan gene (R), which affects white areas only, has been considered distinct from ticking due to T, a dominant. The gene for roan is believed to be dominant to the gene producing clear-white areas in piebalds. The author has had no firsthand experience with the genetics of roaning and, recognizing the difficulty of identifying such a pattern in any but short-haired dogs, prefers to consider the genetic nature of the ground color in Wire-haired Griffons as still uncertain.

The genes in this breed are probably as follows:

$$A^s \quad b \quad C \quad D \quad E \quad g \quad m \quad \underset{\displaystyle s^w?}{\overset{\displaystyle s^p}{|}} \quad T$$

POINTERS

In 1914 the existence of the B (black) and b (liver) pair of alleles and of the E and e pair was recognized in this breed by Little. These pairs determine the basic color varieties in the black-liver, orange-lemon series. The genetic types in these varieties are as follows:

BBEE—black, carrying no recessive types,
BbEE—black, carrying brown (liver),
BBEe—black, carrying lemon (black nose),
BbEe—black, carrying brown, orange (brown nose), and lemon (black nose),

bbEE—liver, carrying no recessive types,
bbEe—liver, carrying orange (brown nose),

BBee—lemon (black nose), carrying no recessive types,
Bbee—lemon (black nose), carrying orange (brown nose),

bbee—orange (brown nose), which can never carry recessive types.

It is admitted that the difference between lemon (black nose) coat color and orange (brown nose) may not be distinct. Since, however, there is often a tendency for black-nosed yellows (B) to be lemon in contradistinction with duller-orange brown-nosed (b) animals, these terms have been used.

These basic colors may appear on different backgrounds of white spotting, either $s^p s^p$ or $s^p s^w$ piebald, usually with 25 per cent or more of the body surface pigmented, or $s^w s^w$, with roughly 20 per cent or less of the body surface pigmented. These figures are an approximation and cannot be taken as definite criteria of the genetic constitution without actual breeding tests.

Many of this breed in the United States will probably prove to be $s^w s^w$, or extreme-piebald spotting. On the other hand, in Europe, especially in England, a considerable incidence of dogs with heavy side spots or saddles is likely. These are usually piebalds of $s^p s^w$ or $s^p s^p$ constitution.

The white areas may be either clear white, indicating a formula of tt (absence of ticking), or they may show different degrees of ticking, being TT or Tt in genetic constitution.

The colored areas are usually uniform in shade, which is indicative of the presence of the A^s (self) gene. Occasionally, however, when the pigment covers the areas where tan pigment is found in tan-point ($a^t a^t$) animals, the presence of that pattern can be detected. It would therefore be possible to obtain clear tricolors if the muzzle and lower limbs were pigmented. These should be most easily recognized in blacks, where the contrast between black and tan is more marked than it is in liver and tan.

The genes found in Pointers include the following:

$$
\begin{array}{ccccccccc}
A^s & B & C & D & E & g & m & s^p & T \\
| & | & & & | & & & |\diagdown s^i? & | \\
a^t & b & & & e & & & s^w & t
\end{array}
$$

POINTERS (GERMAN SHORT–HAIRED)

Like the Wire-haired Pointing Griffons, this breed is always bb brown (liver) in color. There is apparently no record of a B (black) animal. This breed also appears to be consistent in possessing the E gene. There are no available records of ee (yellow or red) animals.

The situation as regards spotting is as follows: solid-colored (S), piebald (s^p), and extreme-white piebald (s^w) animals are described. The solid-colored individuals may be of three genetic types, SS (solid, carrying neither piebald nor extreme-white piebald), Ss^p (solid, carrying piebald), or Ss^w (solid, carrying extreme-white piebald). The spotted types are $s^p s^p$ (piebald, carrying no recessive), $s^p s^w$ (piebald, carrying extreme-white piebald), or $s^w s^w$ (extreme-white piebald, unable to carry anything else).

The ticking factor, T, may or may not be present in any of these types, although its presence could not be visibly detected in any animal with S unless that individual had a white muzzle, chest, or feet, where the ticking could be seen.

It seems probable that a roan gene exists and that some of this breed may show it as a ground color.

The genes found in this breed are as follows:

$$A^s \quad b \quad C \quad D \quad E \quad g \quad m \quad \begin{array}{c} S \\ | \\ s^p \\ | \\ s^w \end{array} \quad \begin{array}{c} T \\ | \\ t? \end{array}$$

RETRIEVER (CHESAPEAKE BAY)

This is another all-brown (bb) breed with black (B) unrecorded. All animals are probably SS in constitution, as they are solid-colored. When any white appears, it is likely to be the result of modifying genes which handicap the expression of S, allowing white markings to occur. This is allowed, but not encouraged, by the Standard.

The recognized incidence of grades of pigment depth from dark brown to "dead grass" or "straw" indicates that some individuals have the gene C, for full pigment depth, while others have the gene c^{ch} (chinchilla), which greatly reduces the depth of brown pigment. Until more extensive data are obtained, it will be satisfactory to classify the types as follows:

CC —deep brown to medium,
Cc^{ch} —medium brown,
$c^{ch}c^{ch}$—light-brown—dead grass, straw.

The genes in this breed are as follows:

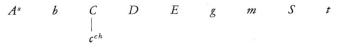

$$A^s \quad b \quad \begin{array}{c} C \\ | \\ c^{ch} \end{array} \quad D \quad E \quad g \quad m \quad S \quad t$$

RETRIEVERS (CURLY–COATED, FLAT–COATED)

These breeds have both black (*B*) and brown (*b*; liver) varieties. There are, therefore, three color types as follows:

BB—black, carrying no recessives,
Bb —black, carrying brown (liver),
bb —brown, carrying no recessives.

All animals are supposed to be solid-colored (*SS*) in constitution. As in other solid-colored breeds, animals with small chest spots occur and are allowed by the Standard. These spots are probably due to a number of modifying genes and are not the same as the true self-perpetuating Irish spotting seen in Basenjis.

The genes in this breed are as follows:

$$A^s \quad B \quad C \quad D \quad E \quad g \quad m \quad S \quad t$$
$$\quad\;\; \mid$$
$$\quad\;\; b$$

RETRIEVERS (GOLDEN)

All records available indicate that this breed uniformly possesses *B* (black) and that *b* (brown) has not been observed. The red-yellow color is apparently due to the action of the *ee* gene combination, so that the normal formula is *BBee*.

There should not be any white spotting in Golden Retrievers. A great majority of them are *SS* in genetic constitution. Mismarking occurs in a number of individuals. Among the co-operators' records at the Jackson Laboratory is one which indicates that the mutation from *S* to *s^i* (Irish spotting) may have occurred at least once. Since the Standard penalizes that amount of white, the Irish-spotted animals will probably not be used for breeding.

Since red coat color "as dark as an Irish Setter" is not desired according to the Standard, most animals are of a considerably lighter shade. They fall within the limits of pigmentation produced by the gene *C* for full-color depth. The author has personally observed one purebred individual of a light dull-lemon yellow, which fact in his experience with other breeds would suggest that the animal was $c^{ch}c^{ch}$, or chinchilla yellow, rather than CC or Cc^{ch}.

The genes in this breed are as follows:

A^s	B	C	D	E	g	m	S	t
		$\mid$		$\mid$			$\mid$	
		c^{ch}		e			$s^i?$	

RETRIEVERS (LABRADOR)

The gene for B (black) is widely distributed throughout the breed. Dr. James Miller of Vancouver, B.C., has written me that liver colored (bb) individuals of this breed are "well known in British Columbia and in Washington and Oregon."

The exact genetic nature of the yellow varieties has not been definitely determined by experimental breeding. It seems likely, however, that they are commonly ee in nature. This is made more probable by the absence of dark areas, or a sable pattern, in the yellow animals.

Burns (1952) reports the crossing of a sable Collie with a yellow Labrador. All-black puppies resulted. This is good evidence that the yellow Labrador is of the A^sA^see type, which when crossed with the a^ya^yEE sable Collie reconstructs A^sE black-coated dogs.

Mismarked individuals have been reported, as in practically all other genetically SS solid-colored breeds. The Standard allows a very small white area on the chest but in no other region. One yellow puppy has been reported with a white spot in which small yellow flecks occurred. This might indicate the presence of the gene T for ticking. One cooperator records two puppies which are clearly black with tan points, a^ta^t in constitution.

The following genes occur in this breed:

A^s	B	C	D	E	g	m	S	t
$\mid$	$\mid$	$\mid$		$\mid$				$\mid$
a^t	$b?$	$c^{ch}?$		e				$T?$

SETTERS (ENGLISH)

Because the liver (*bb*) types of the English Setter, either with or without tan points, are less often seen or recorded than are *Bb* or *BB* types, we shall discuss here only those types which have the ability to form black pigment and are probably *BB* in constitution. Such dogs with *BB* genes are either black with tan points, which with white spotting ($s^p s^p$) makes them tricolor ($a^t a^t$), or they are black and white without tan points ($A^s A^s$ or $A^s a^t$).

They usually show dominant ticking (*T*) or roan in the white areas. Whether or not the roan is due to a large number of closely occurring small ticking spots is a question. In a long-haired dog such coloration could very well appear like roan. The fact that roan *without* some clearly recognizable ticked spots is rarely seen in Setters makes it seem possible that they are different degrees of expression of the same gene. The clear black-and-whites or clear tricolors lack the ticking gene and are *tt* in formula.

In addition to this series of types showing black pigment in the *coat*, there is a series of black-nosed orange-and-lemon animals with or without ticking or roan. These differ from their black-coated counterparts by a single gene *e*, the black-coated series being *E*, which is dominant. No two dogs of the orange-lemon series, when mated together, should give any black or tricolor except in the exceedingly rare event of a mutation.

In the orange-lemon series inspection will not reveal which animals are genetically tan-points ($a^t a^t ee$) and which are $A^s a^t ee$ or $A^s A^s ee$ and hence would fail to show the tan-point pattern if *E* was added. Such animals would be black and white and not tricolor.

The main pure-breeding types are thus:

$A^s A^s EEs^p s^p TT$—black and white, ticked and/or roan,
$A^s A^s EEs^p s^p tt$ —black and white (clear),
$a^t a^t EEs^p s^p TT$ —tricolor, ticked and/or roan,
$a^t a^t EEs^p s^p tt$ —tricolor (clear),

$A^sA^sees^ps^pTT$ —orange and white or lemon and white, ticked
$a^ta^tees^ps^pTT$ and/or roan.[1]

$A^sA^sees^ps^ptt$ —orange and white or lemon and white (clear).[2]
$a^ta^tees^ps^ptt$

A consideration of this variety of genetic types will convince the breeder of the fact that only by extensive, careful records of litters will the genetic formula of any individual dog be determined. The breeder should also remember that, although we have considered the piebald (s^p) gene to be present in all the types listed above, the animals are often obtained in the extreme-white piebald (s^ws^w) form. In these s^ws^w animals side and saddle spots will usually be missing, and pigment other than ticking will ordinarily be confined to the head and/or the areas near the base of the tail. As in Pointers, the separation of the s^p and s^w types by mere observation, without breeding tests, is at best approximate, rather than scientifically accurate.

The following genes are found in this breed:

$$\begin{array}{cccccccccc} A^s & B & C & D & E & g & m & s^p & & T \\ | & | & & & | & & & | & & | \\ a^t & b & & & e & & & s^w & & t \end{array}$$

SETTERS (GORDON)

This is a breed where only one basic color variety is recognized. This is black with tan points. The genetic formula is BBa^ta^tCCEE.

Two possibilities of variation from this type exist. The first may cause a very light straw-tan to appear on the points in place of the usual grades of medium or dark tan. This change in depth of tan is probably sometimes caused by the c^{ch} (chinchilla) allele, which replaces and is recessive to the gene C for

---

[1] In these varieties the A^s and a^t types cannot be distinguished by inspection. All the black-hair pigment, by the distribution of which these patterns are distinguished, is absent in ee animals.

[2] The preceding note applies here also.

full pigmentation. Since the c^{ch} gene does not affect black areas, its effects would only be seen in the tan areas of the coat.

The other departure from the established type is the occasional appearance of all-red animals of the same shades as Irish Setters. This variation is undoubtedly caused by the replacement of the gene E (for extension of dark pigment) by its allele e (which keeps all dark pigment from the coat). The red animals are, therefore, $a^ta^tBBCCee$ in constitution. This means that if two reds were bred together all their progeny would be red, and this variety could be established at once if it was desired.

The following genes are probably present in this breed:

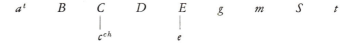

$$a^t \quad B \quad C \quad D \quad E \quad g \quad m \quad S \quad t$$
$$\quad\quad\quad\; | \quad\quad\quad\; |$$
$$\quad\quad\quad c^{ch} \quad\quad\; e$$

SETTERS (IRISH)

This is another one-color breed of the genetic formula A^sA^s-$BBCCee$, giving only reds.

The deep golden chestnut or mahogany-red desired by the Standard is probably the result of a combination of the action of modifying genes for deepening the color and optimum vitamin supply and nutritional conditions, which in many breeds encourage the full expression of color genes.

Although gene S for solid or complete pigmentation is present in a great majority or in all members of this breed, with no clear evidence of the presence of its s^i (Irish-spotting) allele, there are often small areas of white on the chest, toes, throat, nose, or forehead. These are probably SS animals with minus modifiers preventing the full expression of that gene in some areas. In England piebald Irish setters with the gene s^p are recognized and are not discriminated against. This variety is not recognized in the United States.

One other very interesting variation has been reported. Out

of a total of 1,197 pups comprising 129 litters of Irish Setters reported by cooperative breeders, one puppy in a litter of 10 was *solid black*. There seems to be little doubt that this is an authentic case of mutation from *e* (restriction of all-black pigment) to *E* (which allows black pigment to be formed over the whole coat in the presence of gene A^s). Other records of this sort of exception to red would be important and should be reported.

One record from a cooperator describes a single puppy with a black spot on the coat. This variation, which has been observed a number of times in Cocker Spaniels, is discussed in Part Two of this book under the heading of somatic mutations, of which it seems to be an example. A mutation of this kind means that during the development of the puppy the gene *e* in a chromosome of a cell mutated to *E*. Thereafter the descendants of the mutated cell reproduced the *E* condition as faithfully as the unmutated cells produced *e*. The *E*-producing cells as a group eventually proved to be the basis from which an area of the skin with its hair follicles arose. All the hairs from these follicles were *E* in type and therefore produced black pigment instead of red.

In addition to reporting blacks, Burns (1952) has recorded the incidence of the black-and-tan and "shaded sable" color varieties. Marchlewski (1930) has reported blacks but not black and tan. The appearance of black-and-tan and sable types means that the A^s gene, which is usually completely uniform in this breed, has mutated to a^t (tan points) and to a^y sable-tan. Personally I have no records of these varieties in this country.

The following genes are present in this breed:

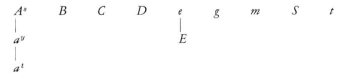

SPANIELS (AMERICAN WATER)

In this breed the Standard requires that all dogs be liver (*bb*) in genetic constitution. No record of any dogs with black (*B*) is obtainable. All animals have the *E* gene and the A^s gene.

Although some individuals have small white areas on the toes and/or chest, it seems probable that they, like the solid-colored animals, are *SS* in formula and that those which show any white do so because of modifying genes and not because of the action of the s^i gene.

The genes in this breed are probably as follows:

$$A^s \quad b \quad C \quad D \quad E \quad g \quad m \quad S \quad t$$

SPANIELS (BRITTANY)

The fact that any black in the coat and a black nose are dis-qualifying characters in this breed means that all animals, whether liver or orange, are really liver (*bb*) in constitution. The gene *B* for black is not present.

The basic formulas for the two standard colors, liver and orange, are, in their simplest form, as follows:

$bbEEs^ps^pTT$—liver piebald with ticking,
$bbEEs^ps^ptt$ —liver piebald without ticking,
$bbees^ps^pTT$ —orange piebald with ticking,
$bbees^ps^ptt$ —orange piebald without ticking.

It is probable that a hereditary type of roan is present, but because of the length of the hair it is not possible to distinguish between heavy ticking on the skin or at the base of the hair and true roan unless careful breeding experiments are carried out.

The following genes are present in this breed:

$$
\begin{array}{ccccccccc}
A^s & b & C & D & e & g & m & s^p & t \\
| & & & & & & & | & | \\
a^t? & & & & & & & s^w? & T
\end{array}
$$

SPANIELS (CLUMBER)

The colored areas in Clumbers are lemon or orange, apparently forming a continuous shaded series from light to dark. This means that they are *ee* in formula. Since dark hazel eyes and dark noses are desirable, it is probable that the *B* (black) factor is present.

Pigment areas are, so far as possible, limited to the head. This fact, combined with the incidence of all-white animals, suggests that the breed is extreme-white piebald, or $s^w s^w$, in constitution.

The following genes are found in this breed:

$$A^s \quad B \quad \underset{\underset{c^{ch}?}{\mid}}{C} \quad D \quad e \quad g \quad m \quad \underset{\underset{s^p?}{\mid}}{s^w} \quad \underset{\underset{T}{\mid}}{t}$$

SPANIELS (COCKER)

This tremendously popular breed is so widely distributed that it is probably not possible to get a complete picture of all the color varieties that may have been, or are being, produced. Coat-color inheritance in Spaniels has been studied for some time by a number of investigators and breeders.[3] As early as 1915 Barrows and Phillips published an analysis of some of the basic color genes, and their work is still helpful.

The difference between blacks, black-and-tans, and all types of black-nosed reds, on the one hand, and livers, liver-and-tans, and all types of brown-nosed reds, on the other, resides in the *B–b* pair of genes. The black series may be *BB*, which does not carry liver, or *Bb* in formula. There is no visible difference between black dogs of these two types. All individuals of the liver series are *bb* and cannot carry black.

For the sake of simplicity we shall confine our discussion of genetic types to those having the gene *B* for black. The reader can transfer to a complete liver (*b*) series the conclusions reached for the black-pigmented types unless otherwise noted. He can visualize the various color patterns in which liver occurs in-

[3] In 1948 the writer published a brief note on genetics in Cockers. Unfortunately, symbols that differ from these now employed were used. It is believed that the present series is more consistent and more acceptable.

stead of black by merely replacing the one dark pigment with the other wherever it occurs on the coat.

Bicolor, i.e., black with tan points, arises from the presence of the gene a^t. As in other breeds having this pattern, there can be great variation in Cockers in both the depth and extent of the tan areas. The extent of the areas seems to depend upon modifying genes. The depth will be discussed under gene C, which affects the depth of red or yellow pigment wherever it occurs.

The usual form of the A series in Cockers is A^s, the gene for a solid-colored coat. This holds true for both the black and the red series. There are, however, a number of apparently authentic records of two red parents producing black puppies. This indicates that two distinct genetic types of red (a^y and e) exist in Cockers. The common type is undoubtedly due to the presence of the ee combination. This is also true of Setters and Spaniels in general. Cockers of this type are A^sA^see. The other way of producing a red coat is by the change of A^s to a^y. This is true in Basenjis, Dachshunds, and Irish Terriers, which are a^ya^yEE. Cockers of this type would not usually be visibly different in appearance from the A^sA^see reds.

Since black coat color results when A^s and E are present together, a mating between an A^sA^see red and an a^ya^yEE red will produce A^sa^yEe individuals, which will be black because A^s and E are both present. Synthetic black dogs of this type have repeatedly been produced from various crosses between breeds at the Laboratory. Sometimes such blacks are a clear, deep black in color. More often they have a peculiar reddish sheen, which appears through the black in certain lights and which may be very noticeable in adults. In some cases reddish hairs may appear in the coat.

Although the evidence for the existence of two types of red cockers is strong, it is interesting that few if any sables or tans which are a^y and which carry black and tan have been reported. One would expect that a^y reds which carried bicolor a^ya^tEE

might show black or dark hairs along the back, at least at birth. If breeders observe such pups, it would be interesting to have them reported.

To understand the relation of black, red, and bicolor in the *ee* type of red Cocker, the reader is referred to the section on English Setters. In this breed no complication is introduced by a second (a^y) type of red, which is probably present in some lines of Cockers. The genetic behavior of the Setters would be the same as that of Cockers having only the *ee* red.

Red Cockers with a small black spot are occasionally reported. These are probably somatic mutations. This topic is discussed in Part Two and also in this part under Irish Setters.

For some time fanciers have felt that red and black Cockers differ from each other in conformation, quality or quantity of coat, and even temperament. The author does not possess any direct data on these points.

It is interesting to note that the amount and distribution of white spotting varies somewhat in the two color types. The reader is referred to the discussion of the S locus in Part Two, where the relationships of spotting patterns are brought out by a study of the various manifestations of the genes at that locus (Irish spotting, piebald spotting, and extreme-white piebald).

The usual parti-colored or tricolored Cocker is genetically piebald ($s^p s^p$). Roughly between 30 and 80 per cent of the coat is pigmented. This is the normal range of expression of the $s^p s^p$ type. The range of piebald spotting in Cockers is very wide and resembles genetically the situation found in Beagles. The reader is referred to the discussion of piebald spotting in Part Two where a chart of Beagles is shown (p. 78).

At the stage of extreme reduction in dark pigment a number of white pups have been recorded. These may be the result of two different and distinct genetic processes. Some of the whites become very light cream or straw-colored with increasing age. These are *extremely pale ee* and/or $a^y a^y$ yellows due to the chinchilla (c^{ch}) or extreme-dilution (c^e) genes. Others may stay clear white or may develop light pigment only on one or both ears. These are probably $s^w s^w$, or extreme-white piebalds, genetically. Some such whites may be blue-eyed. Theoretically there is no reason why both genetic types of white should not occur simultaneously in a single animal.

At the most heavily pigmented end of the various spotted types of Cockers one encounters a great many mismarked animals. These are penalized heavily by bench-show standards, yet they persist in large numbers and crop out occasionally in almost every line or family. It is likely that these mismarked individuals can result from any one of three genetic causes. The most common cause of mismarking is probably the action of minus modifying genes on the expression of gene S for solid coat color. This problem has been discussed in Part Two under the gene S. The sporadic appearance of this type of mismarking cannot be prevented.

The next most frequent type of mismarking is due to incomplete covering up of the piebald or extreme-white piebald pattern when it is carried by animals with the S gene. These Ss^p animals may be solid-colored with no white, but very often they show white on the muzzle, head, chest, or feet. This type may appear *de novo* whenever crosses between solid-colored S and piebald s^p or extreme-white piebald s^w are made.

Many cooperative breeders have reported small white mismarking at birth, which has later disappeared. This may be due to increasing activity of the S gene in Ss^p animals, with the eventual result that the dominance of S becomes complete in that particular region and the white spot disappears.

The third possibility, not yet confirmed by breeding experiments, is that there may be an occasional Cocker which is genetically Irish-spotted ($s^i s^i$). This pattern, acceptable in Basenjis, would result in mismarking in Cockers were it present. It would therefore probably tend to disappear, because the Irish animals would not be used for the breeding of solid-colored varieties and are of no bench-show value under any circumstances.

The only way to determine the genetic type of mismarking is by carefully controlled and observed breeding tests in each individual mating.

Blue dilution (d) has been reported by cooperators in two different cases with a single pup in each case. This variety has not been established, which seems a pity since it should produce a very handsome color type.

In Cockers the following genes are present:

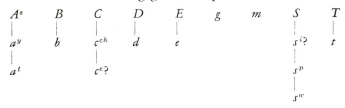

$$A^s \quad B \quad C \quad D \quad E \quad g \quad m \quad S \quad T$$
$$a^y \quad b \quad c^{ch} \quad d \quad e \qquad\qquad s^i? \quad t$$
$$a^t \qquad c^e? \qquad\qquad\qquad\qquad s^p$$
$$\qquad\qquad\qquad\qquad\qquad\qquad\qquad s^w$$

SPANIELS (ENGLISH SPRINGER)

Both black (B) and liver (b) animals occur in this breed. The blacks which carry liver are Bb in constitution and are indistinguishable to the eye from the true-breeding BB individuals. All livers are bb in formula, and two livers bred together will never throw black-pigmented pups unless a mutation occurs.

Both blacks and livers may have tan points ($a^t a^t$), although most of them are without the tan, being A^s. Two dogs with tan points, when bred together, will give only tan-point pups. Usually $A^s a^t$ animals give no visible trace of carrying a^t, but occasionally they will show a reddish tinge in the black or brown pigment located in the tan-point areas.

Practically all Springers are piebald-spotted (s^p). They may show a good deal of pigment and approach the appearance of the various grades of Irish spotting. The great rarity of mismarked animals, however, is an argument against the actual presence of the s^i (Irish) gene in this breed.

Ticking (T) may or may not be present. When it is present, it is often confined to the extremities and muzzle. Burns (1952) has also mentioned this pattern.

In Springers, as in other breeds, the relationship of ticking to roan is complex. Both are certainly inherited, but it is still very uncertain whether they are due to distinct genes, to two alleles of the same gene, or to the grades in expression of ticking, which may be clear-cut at one extreme and the roan intermixture of colored and white hairs at the other.

Red-and-lemon piebald animals, with the lighter pigment on the points, are sometimes, though rarely, seen. They may at times be confused with pale liver with tan points, from which they are, of course, genetically distinct. They differ from black-and-liver piebalds by being *ee*, instead of *Ee*, or *EE*, in genetic constitution.

The common Standard types of the breed indicated by single representation of the genes they show are as follows:

A^sBEs^p—black piebald,

a^tBEs^p —tricolor (black-tan-pointed piebald),

A^sbEs^p —liver piebald,

a^tbEs^p —liver tricolor (liver-tan-pointed piebald),

A^sBes^p —red piebald (the a^te type carries the gene for tan points,
a^tBes^p but since only red pigment is formed there is no visible contrast between the rest of the coat and the various areas in which the red occurs),

A^sbes^p —lemon piebald (the statement concerning the a^te type in
a^tbes^p red piebalds applies here also).

Any or all of the varieties listed may be ticked or clear-spotted. Any or all of them may also show roan, which is possibly due to a special gene dominant to clear spotting. In some cases in animals with hair as long as that of Springers, ticking with very small spots may mimic the appearance of true roan. Breeding tests are necessary to distinguish between the two.

The genes in this breed are as follows:

A^s	B	C	D	E	g	m	s^p	T
a^t	b			e			s''	t

SPANIELS (FIELD)

A great variety of colors are allowed in this breed. There are blacks and livers, both in solid-colored and piebald types. The piebald type is not desirable but it is allowed. There are also mahogany-red, orange, and yellow (both solid and piebald), tan-point, ticked, and roan varieties. Most Field Spaniels have Irish spotting, which is allowed.

The reader will find that the discussion of basic color types under Cocker Spaniels applies to Field Spaniels as well. While some Cockers are piebald, the same fundamental color types in solid colors can be explained for Field Spaniels by the substitution of the gene S for the gene s^p in each formula.

The genes present in this breed are probably as follows:

SPANIELS (IRISH WATER)

This breed, like the Chesapeake Bay Retriever, is a uniform liver (bb) type. All mismarking is discouraged. The genetic formula is $A^sA^sbbCCEESS$.

Because of selection and uniformity of genetic type, there should be little color variation within this breed except in the depth of liver pigmentation. In all probability this variation is due to c^{ch} (chinchilla) replacing C (full depth) in some individuals.

The genes in Irish Water Spaniels are:

SPANIELS (SUSSEX)

The only standard color is "rich, golden liver." Variations, both in lighter- and darker-liver coat color, are regarded as undesirable. The Standard considers these variations as denoting "unmistakably a recent cross with the black, or other variety of the Field Spaniel." While such a cross would explain the observed results in some cases, in others variation in the modifiers of, or in the form of, the C gene itself could produce the same effects with no outcross whatever.

The genes present in this breed are as follows:

$$A^s \quad b \quad C \quad D \quad E \quad g \quad m \quad S \quad t$$

SPANIELS (WELSH SPRINGER)

In this breed the allowed color is "dark rich red and white." As the eyes are hazel or dark and the nose brown or flesh, it is probable that gene b, for liver pigmentation, is present in all individuals. The red color is undoubtedly due to the ee genes, as is the case in almost all Spaniels and Setters. Piebald spotting ($s^p s^p$) is characteristic of the breed. Ticking (T) may or may not be present, as in English Springers.

This would make the formula of the two color types $A^s A^s bbCCees^p s^p TT$ or Tt (red-piebald, ticked) or $A^s A^s bbCCees^p s^p tt$ (red-piebald, unticked).

The following genes are probably present in this breed:

$$A^s \quad b \quad C \quad D \quad e \quad g \quad m \quad s^p \quad \overset{T}{\underset{t}{\mid}}$$

WEIMARANERS

Observation of individuals of this breed at various bench shows indicates that there is one basic genetic type with respect to genes for fundamental color. The eye and coat color combine to give strong evidence that all animals are bb brown, with no gene B for black. The flat, dull quality of the coat color is due to the fact that all individuals are dilute or dd in formula. This is the only breed in which the universality of *dd* is characteristic.

Rigid limitations of any white to a small chest spot indicates that *SS*, rather than other spotting genes, is the genetic formula for the breed.

Discrimination against any yellow tinge on the chest shows that, if the a^t (tan-point) pattern has occurred, it has been eliminated by selection, thus leaving the breed A^sA^s.

While the deep-colored individuals are undoubtedly *CC* in formula, the appearance of the most lightly pigmented animals suggests that some are $c^{ch}c^{ch}$ or Cc^{ch}.

We may, therefore, list the prevalent types as follows:

$A^sA^sbbCCddEE$ —dark-colored,
$A^sA^sbbCc^{ch}ddEE$ —medium-colored,
$A^sA^sbbc^{ch}c^{ch}ddEE$—light-colored.

The genes in this breed are as follows:

$$A^s \quad b \quad C \quad d \quad E \quad g \quad m \quad S \quad t$$
$$\quad\quad\quad\quad |$$
$$\quad\quad\quad c^{ch}$$

CHAPTER XI

Sporting Dogs (Hounds)

AFGHAN HOUNDS

LACK of firsthand contact with sufficient numbers of this breed and the wide variation in terminology used by fanciers in describing the many color types produced combine to make a discussion of the genetic behavior of these types necessarily tentative and subject to revision when more extensive and accurate data are available. It will be best to proceed from what seems to be relatively consistent and well established to the less certain and more debatable color combinations.

Discrimination by the Standard against white markings undoubtedly means that the breed is self or solid-colored, SS in formula. Mismarked individuals would, in that case, be those in which modifying factors had prevented the spread of pigment over the complete body surface, thus allowing small white areas to appear on the feet, chest, muzzle, or forehead. The presence of solid-black individuals means that the genes A^s and E are both present, while the incidence of black with tan points is evidence that the $a^t a^t$ combination also exists.

Leaving the question of masks for later discussion, it seems probable that two types of red or fawn may occur. The existence of the ee type, similar to the ordinary red series in Setters and Pointers, has been established by a cross made at the Jackson Laboratory. Cases of dark-sable pups born with a good deal of black or brown dorsal or lateral pigment have been reported and seen. They usually become lighter with age and may eventually possess a clear-red or tan coat. This behavior is characteristic of reds produced by the a^y gene either in $a^y a^y$ or $a^y a^t$ individuals. Black-and-tan pups from a mating of two reds have also been reported. This would suggest the presence of an a^y type of red or tan. For the present, then, it will be well to admit the possibility of both e and a^y reds in Afghans.

The presence or absence of a mask, while undoubtedly due to genetic influences, has not yet been clearly analyzed in an adequate series of controlled matings. It is probable that the extent and depth of pigmentation of the mask vary considerably with the ground color and genetic constitution. This naturally complicates its study as an inherited character and has led to the employment of a number of descriptive terms which may mean one thing to one breeder and something quite different to another. In general, it seems that the presence of a mask is dominant over its absence.

Cream, straw-colored, or even white individuals occur and are probably due to the action of $c^{ch}c^{ch}$ instead of C, the factor for complete pigmentation. Yellow animals of ee constitution are diluted to cream color or white by c^{ch}. Tan animals which are a^y may also be made much lighter by that gene. Its presence in creams or whites may also reduce the total pigment so much as to make the contrast between a mask and the rest of the animal less distinct or nonexistent. Blue dilution due to the action of gene d has also been reported but is rare. Combination of the $a^y a^t c^{ch} c^{ch}$ or $A^s c^{ch}$ and dd genes may well produce the variety called silver.

While Afghan breeders speak of brindling and grizzling, it is doubtful whether these colors are anything more definite than variations of the sable pattern. The ability of this pattern to produce areas of mixed light and dark hairs is well known.

The following genes are present in this breed:

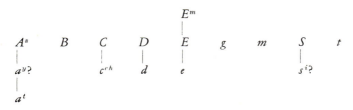

BASENJIS

All of the members of this breed have the Irish ($s^i s^i$) type of white spotting. This consists of more or less white on the feet, legs, chest, and tail tip with, at times, extension of the white to cover more of the ventral surface and to form a white collar of varying width.

The usual coat color is red. The pups may be almost clear red from birth but are more apt to be grizzled, especially on the middorsal line and on the sides, becoming clear red as adults. Repeated breeding tests have shown that the red coat color is due to the a^y gene and that a great majority of the animals are $a^y a^y$ in formula. Although there is some variation in the depth of the red pigment, it is probably due to modifiers of gene C for full pigmentation rather than to any other form of that gene, such as c^{ch}.

Blacks with tan points ($a^t a^t$) are also recorded. The fact that blacks are allowed by the Standard suggests that the A^s gene may have been present, but in this country, at least, it has been eliminated by selective breeding.

The blue and the cream coat color reported occasionally may well be due to the very rare occurrence of the d gene. These varieties are *not* preferred.

The genes present in this breed are as follows:

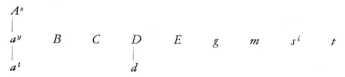

BASSET HOUNDS

Basset hounds may be solid-colored with no white or either tan ($a^y a^y$ or $a^y a^t$) or black with tan-point ($a^t a^t$) patterns. These colors may also occur in definitely piebald patterns, tan and white or tricolor. Most individuals seen by the author have had black (B), rather than liver (b), pigment.

It is probable that the solid-colored varieties (S) which carry piebald (s^p) as a recessive, and are therefore Ss^p, will show white on the feet, chest, or muzzle, mimicking the Irish pattern in appearance.

Some Bassets show ticking (T); others do not. In those that have ticking, it may or may not be confined to the muzzle, chest, and legs.

The basic varieties would then be:

$a^t a^t BBSS$	—black, with tan points,
$a^y a^t$ or $a^y a^y BBSS$	—tan,
$a^t a^t BBSs^p$	—black, with tan points and Irish-type spotting,
$a^y a^t$ or $a^y a^y Ss^p$	—tan, with Irish-type spotting,
$a^t a^t BBs^p s^p$	—tricolor,
$a^y a^t$ or $a^y a^y s^p s^p$	—tan-and-white piebald.

The depth of tan pigment may be greatly reduced and lightened by the action of $c^{ch} c^{ch}$ instead of CC, the gene for full-pigment depth.

This breed has the following genes:

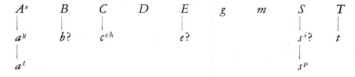

$$
\begin{array}{ccccccccc}
A^s & B & C & D & E & g & m & S & T \\
| & | & | & | & | & & & | & | \\
a^y & b? & c^{ch} & & e? & & & s^i? & t \\
| & & & & & & & | & \\
a^t & & & & & & & s^p &
\end{array}
$$

BEAGLES

In the A series this breed is usually $a^t a^t$ (dark saddle with tan-points). The extent of tan is usually considerable. In 149 litters of $a^t \times a^t$ individuals, a total of 729 pups, all a^t, was recorded. The existence of animals with any other member of the A series is unusual. An analysis of 160 litters reported on blanks from cooperative breeders indicates that the A^s or solid-black distribution of pigment, with no tan areas, was recorded in two litters. In these litters one parent was tan-pointed (a^t), the other was black (A^s). It is probable that in both cases the black parent was $A^s A^s$ in formula, for all 14 pups were black, with no tan recorded, and therefore were $A^s a^t$ in constitution. The unfortunate habit of some careless breeders of calling a black-

and-tan with a small amount of tan "black" may account for these exceptions rather than mutation to the A^s gene. Whether the a^y member of the series has been recorded is doubtful, although the fact that sable (a^y) dogs at times appear to be grayish at birth and that 3 "gray" pups are recorded in one litter which also contained 4 typical tan-points indicates that breeders should be on the alert for sable pups with no black saddle.

Most Beagles are *BB* in formula and give only pups in which the dark pigment is black. In the total of 149 litters derived from matings in which the parents had black pigment, four contained liver pups. In these four litters there was a total of 11 black pups and 8 liver pups, indicating that the parents were *Bb* in formula. One mating, presumably of *BB* (black) × *bb* (liver), gave 6 pups all with black pigment; three matings of B*b* × *bb* gave 8 black and 8 liver pups, the exact Mendelian expectation.

In the *C* series, almost all Beagles are certainly *CC* (fully pigmented). The only evidence for the presence of another member of this series is found in the occurrence of individuals in which the tan-colored areas are a pale, flat buff instead of the usual rich tan. The dark areas of the coat of such animals show no modification. Such a situation might be expected if the chinchilla ($c^{ch}c^{ch}$) combination had been obtained. Further studies of the results derived from crosses involving these lightly pigmented animals should be made.

Beagles are usually *DD* (intensely pigmented), but in 3 of 160 litters blue individuals, which must be *dd*, are reported. These animals have slate noses and light eyes. Care should be taken to avoid confusion with the "blue" that results from roan, which is caused by an entirely different genetic background from "blue" dilution (*d*).

In the *E* series it seems certain that a mutation from *E* to *e* has occurred and has given rise to yellow or tan coat color in which no dark pigment is observed. In 149 matings of *E* × *E*

animals 742 pups had extended dark pigment and 30 were *ee*
(tan) with no dark pigment. In five matings of *EE* or *Ee* × *ee*
individuals 16 *Ee* and 8 *ee* pups were produced. One mating of
ee × *ee* gave, as expected, only *ee* pups (4).

The usual Beagle is piebald-spotted (s^p), although Irish spotting (s^i) and extreme piebald (s^w) are also seen. If animals with
no white should appear, they would be either true self-colored
(S) and would segregate clearly in inheritance, or they would
represent the extreme of heavy pigmentation of the Irish (s^i)
pattern and not tend to reappear.

Piebald individuals may be uniform for that pattern ($s^p s^p$) or
may carry extreme piebald and be $s^p s^w$ in genetic constitution.
Reference to Fig. 7 (p. 78) on Beagles will give some idea of
the range of the spotting pattern. In this figure grades 9 and 10
represent the degree of spotting usually shown by $s^w s^w$ animals.
Grades 3–8 inclusive are characteristic of $s^p s^p$ or $s^p s^w$ piebalds
and $s^i s^p$; grades 1 and 2 are within the established range of the
Irish pattern, $s^i s^i$. Use of these grades for purposes of classification does not make the judge infallible. They are merely a
convenience and an approximation and should be so considered.

All Beagles so far recorded have been free from the merle (M)
gene and are nonmerle or *mm* in formula. All of them also appear to be free from the graying gene (G) and are therefore *gg*
in constitution.

The *T* gene for ticking appears to be present in many individuals, although at present no adequate numerical data from
controlled matings are available.

A summary of the genetic situation in Beagles' coat color
may be made as follows:

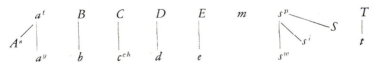

A^s is rare and may actually be a^t, with darkening modifying
factors.

Some comparisons of the more common types are as follows:

1. $a^tBCDEms^p$ —tricolor (black blanket, tan-points),
2. $a^tbCDEms^p$ —tricolor (liver blanket, tan-points),
3. $a^tBc^{ch}DEms^p$—tricolor (black blanket, buff-points),
4. $a^tbc^{ch}DEms^p$ —tricolor (liver blanket, buff-points),
5. $a^tBCdEms^p$ —tricolor (blue blanket, dull-yellow points).

All these forms can be turned at once into tan, yellow, or buff-and-white animals by changing E to e. Types 1 and 3 would have black noses, types 2 and 4 brown or pink noses. The tan in types 1 and 2 would be deep and rich, while in 3 and 4 it would be pale buff. In type 5 it would be a medium yellow with a peculiar flat, lusterless tone. Dogs of type 5 can be expected to have slate-colored noses and light eyes.

The amount of the body surface in all of these forms which will be pigmented will be determined as follows:

S —all the surface,
s^i—white on feet, legs, and chest, probable blaze,
s^p—the usual distribution of colored head and blanket, rest of surface being white,
s^{w}—few colored areas, usually the ears and near base of the tail.

Since there is a great deal of variation in the piebald pattern, the only way in which the genetic make-up of any individual can be accurately determined is by observation of carefully controlled and observed individual matings.

BLOODHOUNDS

Little white spotting is seen in this breed, and it is probable that any white spots observed are the results of modifying genes acting on the gene S for self or solid-colored coat. This breed usually has the factor B for black pigment. Liver (b) and blue (d) have been reported but are not common.

The three color varieties mentioned in the Standard are "black and tan" (a^ta^t), "red and tan" (a^ya^t), and "tawny" (a^ya^y). Puppies born red and tan may become clear tan or tawny with increasing age. The a^ta^t type is "tan-pointed."

If tans appear with appreciable frequency from two black-and-tan parents, these tans would probably be *ee* in formula. The only other way in which they could be produced would be by mutation from a^t to a^y. This would be a great rarity.

There is some variation in the depth of the tan pigment, which may become light yellow or almost straw-colored in some individuals, either because of lightening modifiers or to the change of *C* (the gene for full pigmentation) to its paler form c^{ch}.

The genes present in this breed are as follows:

$$
\begin{array}{ccccccccc}
a^y & B & C & D & E & g & m & S & t \\
| & | & | & | & | & & & & \\
a^t & b? & c^{ch} & d? & e? & & & &
\end{array}
$$

DACHSHUNDS

The red in Dachshunds is usually due to the action of the a^y gene. This is the genetic type of red seen in Basenjis, Irish Terriers, and Bloodhounds. When the red animal carries a^t (tan-points) as a recessive, the color may become darker and a band of dark pigment may cover the middorsal line and sides. These animals are really dark sables, which can be clearly recognized in the long-haired variety, where such individuals are much the same in color as Collies of a similar genetic type.

The basic color varieties are, therefore:

$a^y a^y BB$—clear red, black nose,
$a^y a^t BB$—dark red-sable, black nose,
$a^t a^t BB$—black, with tan-points.

The gene for black (*B*) can also be replaced by that for brown (liver), in which case the three types are as follows:

$a^y a^y bb$—clear light red or yellow, brown nose,
$a^y a^t bb$—clear red, brown nose,
$a^t a^t bb$—liver with tan-points.

In the rough-haired varieties the effect of a^t on a^y, when carried as a recessive, is often marked. Many dark and grizzled animals with tan-points represent the most heavily pigmented individuals of the $a^y a^t$ type.

There is still some question whether a wild-type coat color occurs in rough-coated Dachshunds. Superficially some of these animals look very much like the "pepper and salt" coat pattern of Schnauzers, but adequate evidence is lacking on this point at present. It would be an interesting problem to investigate.

There is also a possibility that brindle (e^{br}) occurs. The pattern has been observed and is illustrated by Reichenbach's *Der Hund* (1836). It is also clearly shown as "*Gestrometer*" by Engelmann (1925, p. 101). Burns (1952) refers to the "rare brindle." In the cooperators' blanks received at Bar Harbor only one puppy is listed as "brindle." A detailed description obtained by correspondence makes it seem probable that it was an authentic case. The writer has not seen this pattern and doubts very much whether it is today more than a rarity.

As far as ordinary piebald spotting is concerned, Dachshunds are all genetically self or SS in formula. The occasional white chest spot, toe tip, or belly spot is due to modifiers which effect the development of the S gene.

The depth of tan pigment varies in Dachshunds, as in most hound breeds. This variation may be caused by modifying factors acting on the C gene for degree of pigment depth. It is also possible that the C gene has changed to its pale form, c^{ch}. The probability that this is the case is strengthened by the existence of considerable variation in the depth of brown (liver) pigment from a very dark chocolate-brown to forms as pale as those in the average Chesapeake Bay Retriever. In such pale animals the tan on the tan points is correlated in depth with that of the basic brown coat color, becoming darker as the brown deepens and lighter as it pales.

Correspondence from two breeders indicated that the "d" mutation (dilute) producing "blue" individuals with pale tan points (a^t) has occurred in this breed. This gene should also affect the

A^y coat colors producing undesirable "washed out" red or yellow.

Dachshunds are also one of the few breeds in which the merle or dappled gene (M) may be present. This gene is dominant over the nonmerle basic color varieties already described. The merle pattern consists of an irregular patchwork of two contrasting colors, as follows:

Ground color	Contrasting color
Black	Blue-gray (sometimes with a reddish tinge)
Liver	Pinkish beige
Deep red	Bright, often light, red

Interestingly, when the merle pattern occurs on tan $a^y a^y$ or $a^y a^t$ animals, it is usually more noticeable at birth and it often disappears entirely with increasing age. In such dappled tans there may, in some cases, be a blue or blue-and-brown-mottled eye or even a small wedge-shaped sector of blue remaining in the iris as the only trace of the merle pattern. A similar condition has been described in Collies and Shetland Sheep Dogs by Mitchell (1935) and others.

The merle pattern is also apt to increase the tendency for white spots to appear on the coat. These have no evident relationship to white spotting formed by any of the piebald (S) series of genes. Even merles (Mm) carrying nonmerle mm may show such spots.

In the pure-breeding merles (MM) produced and observed at the Jackson Laboratory the white areas may cover as much as 80 per cent of the coat, but such animals are not usually pure white as in Collies. Reduction in eye size and in vision may also appear in these MM individuals. As yet the evidence is inconclusive as to the existence of deafness, which is often seen in white MM Collies or Shelties.

A coat-color gene of the M type is a very interesting one. It has distinctly harmful effects when it is not balanced by the normal nonmerle gene m. Thus it could not exist long in a state of nature in competition with the nonmerle type.

It is possible that the production of reds and yellows by the change of *E* to *e* has occurred in Dachshunds. The best evidence for this would be obtaining clear yellows or reds from carefully controlled black-tan-pointed or liver-tan-pointed matings with no possibility of confusion with the ordinary tan (a^y) type.

Breeders should be on the lookout for the appearance of solid-black or solid-liver dogs without tan points or for true brindles, which could theoretically appear by extremely infrequent mutations (or sudden genetic change) from a^y to A^s in the former case or from *E* to e^{br} in the latter. With the great number of Dachshunds being produced, one or both of these new types can be expected eventually, either by mutation or from a line in which they have been masked or hidden for a number of generations.

The genes found in this breed are as follows:

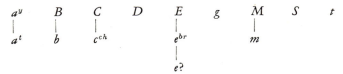

$$
\begin{array}{ccccccccc}
a^y & B & C & D & E & g & M & S & t \\
| & | & | & & | & & | & & \\
a^t & b & c^{ch} & & e^{br} & & m & & \\
& & & & | & & & & \\
& & & & e? & & & &
\end{array}
$$

DEERHOUNDS (SCOTTISH)

The colors normally found in this breed are variations of the brindle pattern. This pattern is one of the most variable found in dogs. It can manifest itself as a few light hairs on a dull-gray or deep blue-gray background at one end of its expression. At the other extreme may be found yellow or sandy-red individuals with a few dark hairs (either solid-colored or banded with dark tips) interspersed in the coat. Between the two extremes are all sorts of striped or stippled types with light areas consisting of almost white or straw to rich tan hairs and with dark areas running through various shades of red to dark gray approaching black. Thus there are dark or light brindles according to the depth, richness, and color of the background.

The foregoing remarks apply to the surface appearance of the pattern. It should be recalled that one attribute of the brindle

(e^{br}) pattern is the presence of alternating light and dark stripes in the undercoat. These may sometimes be observed in Deerhounds at birth or if the outer coat is clipped short.

Modifying factors probably determine the usual range of variation of the pattern. The genetic make-up of these individuals is $BBa^ya^ye^{br}e^{br}CC$ in the darkest type of brindle. A lighter shade of brindle is $BBa^ya^ye^{br}e^{br}Cc^{ch}$. Still lighter, but clearly brindled, is $BBa^ya^ye^{br}e^{br}c^{ch}c^{ch}$. The lightest true brindles are probably $BBa^ya^ye^{br}ec^{ch}c^{ch}$. Finally, wheaten individuals may be formed by a complete change of e^{br} to e; they would be of the formula BBa^ya^yeeCC, Cc^{ch}, or $c^{ch}c^{ch}$. It would follow from this that there would be wheatens of different shades, from deep sandy red to cream, depending upon the C and c^{ch} genes. Also, when two wheatens are bred together, only wheatens should be produced.

The fact that the Standard mentions "blue nose" suggests that some of the lighter individuals may actually be dd or dilute in coat color.

Black masks are sometimes seen. It is uncertain whether they are due to a gene in the E series or to an independent gene. The reader is referred to Part Two for a discussion of this pattern.

Occasionally the Irish spotting pattern s^i occurs. For the most part the coat is self S (without white).

The genes found in this breed are as follows:

$$
\begin{array}{cccccccccc}
 & & & & E^m & & & & & \\
 & & & & | & & & & & \\
A^s & B & C & D & E & g & m & S & t \\
| & & | & | & | & & & | & \\
a^y & & c^{ch} & d? & e^{br} & & & s^i & \\
 & & & & | & & & & \\
 & & & & e & & & & \\
\end{array}
$$

FOXHOUNDS (AMERICAN AND ENGLISH)

The coat color in both American and English Foxhounds is similar to that of the types described in the section on Beagles, to which the reader is referred. In American Foxhounds solid tan or solid black and tan occur fairly frequently. The presence of solid-colored individuals is the result of gene S, and this, in turn, allows the formation of animals with small amounts of white on the feet, chest, or head. These mimic both the more heavily pigmented types of Irish spotting (s^is^i) and the spotting which appears when the animal is Ss^p or Ss^w in constitution.

The genes found in this breed are as follows:

GREYHOUNDS

Whether or not the extreme antiquity of the Greyhound breed has permitted a large number of hereditary color variations to occur or whether crosses with other breeds has introduced new genes is uncertain, but an unusually wide range of coat-color genes has been described in the Greyhound breed. The Standard of the breed washes its hands of responsibility when it says that the coat color is "immaterial." Warren (1927), using data from Volumes 11 to 19 of the Greyhound Stud Book, has very helpfully tabulated various types of matings between different coat-color types.

The presence of both solid-colored blacks and those with the Irish variety of spotting is clearly established. Crosses between blacks produce that color and also brindles, fawns, and reds. Fawn × fawn does not usually produce blacks and only very rarely brindles. When brindles *are* recorded from fawn × fawn matings, it is probable that one of the fawn parents was really a brindle but was so light that the dark hairs of the brindle pattern were overlooked.

It seems certain that there are two types of animals in the fawn-red color varieties. These are probably ee and a^ya^y. Crosses between a^ya^yEE reds and A^sA^see fawns will recreate A^sa^yEe blacks.

Warren's data from red × red or red × fawn matings gave 5 blacks, 14 brindles, 58 fawns, and 78 reds. The blacks and brindles would not have appeared had red-fawns all been of one basic genetic type.

His figures on brindle $a^y a^y e^{br} e^{br}$ or $a^y a^y e^{br} e$ × red $a^y a^y EE$ or $a^y a^y Ee$ are interesting and confirm the genetic explanation of the coat color of both. Brindles, reds, and fawns are the three types expected, and the numbers observed were 108, 31, and 19, respectively.

As for the spotted varieties, S (solid-colored), s^i (Irish), s^p (piebald), and s^w (extreme-piebald) are probably present. There has been no particular selection for or against any of these varieties, and, as a result, they have all been preserved.

The depth of pigment is apparently affected by two genetic influences. One of these is the paling gene of the general color factor C which is symbolized by c^{ch}. This does not affect black animals but makes fawns and especially reds very light, even to a cream or straw color.

The other gene affecting pigment depth is d, for Maltese-blue dilution. Black animals become this color when they are also dd in constitution. Fawns and reds need not be particularly lightened by d, but they take on a peculiar flat, washed-out quality of color at whatever total pigment depth they develop.

This breed has the following genes:

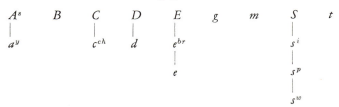

NORWEGIAN ELKHOUNDS

The light pigment in this breed is very pale, sometimes almost white. This suggests that c^{ch}, rather than the gene C for deep pigmentation, is the rule. While a vast majority of these dogs have black (*B*) rather than brown (*b*) as their dark pigment, there is at the Jackson Laboratory a pedigreed and registered bitch which is liver (*bb*) in genetic constitution. In all other respects her coat color is similar to the Standard type. Liver Elkhounds (called "red" by breeders) are apparently common in Scandinavia.

The pattern of dark and light pigment distribution in this breed has not yet been definitely analyzed or classified genetically. It does not appear to be identical with sable a^y or with brindle e^{br}, although in common with a^y more light pigment appears with increasing age.

Elkhounds are born black with a scattering of gray hairs, while the darkest sables that have been recorded in any litter at the Jackson Laboratory are definitely of mixed color with dark and light hairs interspersed and with a tendency for the areas which are tan in black-and-tan (tan-pointed) animals to have a larger proportion of light hairs than do other areas of the body.

For the present and until more firsthand data on crosses involving Elkhounds are available, the coat color which they commonly develop has been classed as agouti or wild color due to a^w, a gene in the A^s, a^y, a^t series. Practically without exception Elkhounds have well-developed black masks.

All Elkhounds are *SS*, or solid-coated. Whenever white areas appear on toes or chest, a condition discouraged by the Standard, they are due to modifying factors acting on the *S* gene.

The genes present in this breed are as follows:

$$a^w \quad B \quad c^{ch} \quad D \quad E^m \quad g \quad m \quad S \quad t$$
$$ b$$

SALUKIS

This is another breed with many recognized color varieties. However, "grizzles" undoubtedly represent dark sable rather than brindle. The Standard speaks of a grizzle and tan, which would be the natural pattern if sable was darkened a degree and the tan-point areas stood out in contrast. The two types of red seen in Greyhounds apparently occur also in Salukis.

White animals are recorded; these may be *very* pale cream-colored individuals, with either one or a combination of the two types of red, plus the paling factor c^{ch}, which reduces the pigment to such a degree that the coat appears to be unpigmented or white. The other possibility is that the white Saluki is a form of extreme-white piebald spotting. Relatively few bench-show Salukis in this country are even piebald; hence whites are probably produced by c^{ch}. Irish spotting (s^i) is frequently encountered in this breed.

The black variety with tan-points is often seen, and the tan varies in depth as one would expect if both factors C for deep pigment and c^{ch} for paleness were present in the breed. Solid blacks due to A^s and tricolors which are probably black-and-tans with the Irish type of white spotting (s^i or s^p) are rarely reported in this country. They are more common in England.

Ticking due to gene T is also found in this breed, an interesting fact when one considers its absence or extremely rare incidence in Greyhounds, Whippets, and Afghan Hounds.

The genes found in Salukis are as follows:

WHIPPETS

In this breed, as in Greyhounds, the Standard sets the color requirements with the one word "immaterial." Actually all the colors described for Greyhounds are found, including black, red, fawn, blue, brindle, cream, and white. Reds, fawns, and brindles are of many degrees of depth of color. They may also occur with white spotting, varying from Irish through piebald to extreme-white piebald or, as mentioned above, to white itself.

The cumulative effects of blue dilution (d) and the paling factor (c^{ch}) produce both light blues and dull, flat-colored smoky-fawns, with bluish-gray eyes and noses, and creams that might even be classed as white.

Although tan Whippets are commonly seen, those that are black or blue *with tan-points* are rare or absent. This means that while the genes A^s for black solid color and a^y for tan are frequently present, a^t is not.

Histories of the origin of the Whippet breed mention a cross or crosses between a Terrier breed or Terrier breeds with the Greyhound. There are also intimations that the Italian Greyhound is involved in its ancestry. At present, Whippets are essentially small Greyhounds in form and in all their coat-color inheritance, except for the comparative rarity of brindles of all types.

Whippets have the following genes:

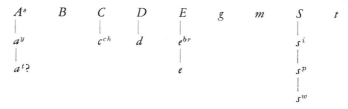

WOLFHOUNDS (IRISH)

Most animals of this breed, like Scottish Deerhounds, are some type of brindle (e^{br}). Wheaten or yellow individuals probably due to e are also found.

White dogs with dark eyes and noses have been reported. Because of the fact that the only type of spotting encountered in this breed is the type of marking which results from modifiers acting on S (the gene for a solid coat), pure-white individuals are not likely to represent an extreme type of spotting. Pure white is probably a pale yellow with pigment so greatly reduced by c^{ch} or a more extreme paling gene c^e that it is not recognizable to the eye.

The discussion of the brindle pattern under the section on Scottish Deerhounds is applicable to Irish Wolfhounds as well. The genes in this breed are as follows:

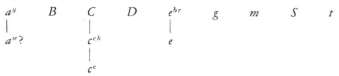

$$
\begin{array}{ccccccccc}
a^u & B & C & D & e^{br} & g & m & S & t \\
| & & | & & | & & & & \\
a^w? & & c^{ch} & & e & & & & \\
& & | & & & & & & \\
& & c^e & & & & & &
\end{array}
$$

WOLFHOUNDS, RUSSIAN (BORZOI)

The fact that the Standard says that white usually predominates in the coat pattern of this breed, combined with the incidence of all-white individuals, indicates that the extreme-white piebald gene (s^w) is certainly present. As solid-colored individuals occur occasionally and ordinary piebalds are frequently seen, the former more especially in England, S and s^p must also be present. Although the Irish type of spotting also occurs, it is not frequently seen.

The tan-point pattern (a^t) producing tricolors in combination with piebald is a common type. The listing of brindles and grays suggests that in this long-coated breed the a^u, or sable-tan, pattern is involved. It seems doubtful that the brindle spoken of in Borzoi is similar to that seen in Irish Wolfhounds or Great Danes. It is much more likely to be dark grizzled sable, which possibly carries a^t (the tan-point pattern) as a recessive coat-color character.

The lemon-and-tan series probably represents variations in depth of the $a^u a^u$ types. Since it is not uncommon to find tricolors with very pale tan areas, it is likely that the c^{ch} gene for producing genetically pale colors may be present in such animals.

Borzoi have the following genes:

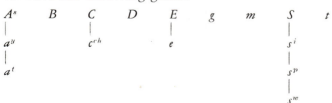

$$
\begin{array}{ccccccccc}
A^s & B & C & D & E & g & m & S & t \\
| & & | & & | & & & | & \\
a^u & & c^{ch} & & e & & & s^i & \\
| & & & & & & & | & \\
a^t & & & & & & & s^p & \\
& & & & & & & | & \\
& & & & & & & s^w &
\end{array}
$$

CHAPTER XII

Working Dogs

BOXERS

THE two common colors in this breed are brindle and fawn. All individuals have black pigment but no liver color has been reported.

A more or less extensive black mask is usually present. Tans or brindles without the mask are sometimes produced, but they are not allowed by the Standard. Indications are that the mask is dominant over no mask. However, this is not yet certain, for the mask varies so greatly in extent that identification of those with a very slight degree of mask development might conceivably be difficult or impossible. As a result, the records on masks obtained from breeders need to be supplemented by those from a series of carefully planned and observed laboratory matings.

The same indefinite boundaries are encountered in the variation of the brindle pattern, which in its least extensive development may easily be overlooked, with the result that the dog is described as tan or fawn. In its darkest type the coat color might easily be confused with black.

More experiments are needed before one can say with certainty whether the tans and fawns are of the $a^y a^y$ or ee type. The absence of black with tan-points ($a^t a^t$) prevents the formation of the $a^y a^t$ combination. If this combination were obtained, it would be a medium or dark sable and would be dominant to $a^t a^t$, black with tan-points. If the ee combination produced the tans and fawns, a mating of a tan Boxer with a black-and-tan such as a Doberman should produce only black-and-tans. If the tans and fawns were $a^y a^y$, such a mating should produce all tans or sable. This hypothesis appears to be correct.

In both brindle and tan types a small amount of white spotting is allowed. Since the Standard restricts the total amount of white to one-third of the body area, the Irish pattern rather than piebald spotting will be selected.

The tabulation of the amount of white spotting in a large population of Boxer pups given in the discussion of the gene S (p. 87) is of interest. It indicates definitely that the gene S for completely pigmented coat is present, as are two other genes affecting spotting. One of these is s^i, the gene which produces Irish spotting, where white is confined to any or all of the following locations: feet, legs, brisket, chest, throat, muzzle, and forehead.

The other spotting gene, s^w (extreme-white piebald), produces all-white or nearly white pups with a little pigment on the head or ears and occasionally at the base of the tail. Since these animals are promptly discarded because they are not approved by the Standard, the white or nearly white Boxer has not been established as a recognized variety. Apparently white was allowed until approximately 1938.

This elimination of whites seems a pity, for they could be a handsome and appealing variety of the breed. As most strains carry the s^w gene, failure to recognize it as a desirable type necessitates discarding a large number of otherwise acceptable puppies.

Genes in this breed are as follows:

$$
\begin{array}{ccccccccc}
 & & & & E^m & & & & \\
 & & & & | & & & & \\
a^y & B & C & D & E & g & m & S & t \\
 & & & & | & & & | & \\
 & & & & e^{br} & & & s^i & \\
 & & & & | & & & | & \\
 & & & & e? & & & s^w & \\
\end{array}
$$

BULL–MASTIFFS

The basic colors of this breed appear to be identical with those of Boxers, namely, brindle and any shade of fawn. The black mask, so essential for bench-show specimens of Boxers, is preferred in Bull-Mastiffs also but is not yet universal in this breed.

Any white, even on the feet, muzzle, or chest, is rigidly disapproved. Constant effort is exerted to obtain only solid-colored dogs.

The discussion of variation within the brindle and fawn patterns of Boxers applies to the Bull-Mastiff as well and should be read by breeders of the latter.

The genes in Bull-Mastiffs are probably as follows:

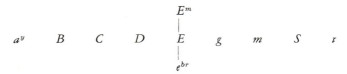

$$a^y \quad B \quad C \quad D \quad \overset{\displaystyle E^m}{\underset{\displaystyle e^{br}}{\overset{\displaystyle |}{\underset{\displaystyle |}{E}}}} \quad g \quad m \quad S \quad t$$

COLLIES (ROUGH AND SMOOTH)

Although variations in the length of the coat may give a slightly different optical effect to certain coat-color patterns in the two varieties of Collie, the genes underlying coat color are the same in both. For this reason the two varieties will be considered and discussed together. The reader will find a good analysis of coat-color inheritance in this breed in an article by Mitchell (1935).

Collies occur in four visibly different color types, all of which have the ability to produce black (B) rather than liver (b) pigment. Individuals may occur in either the sable ($a^y a^y$ or $a^y a^t$) or the tan-point ($a^t a^t$) pattern. In general, the darker or shaded forms of sable will be those which are $a^y a^t$, while the great majority of the clear sables will be $a^y a^y$.

The merle pattern (M) in Collies is apparently due to the

same gene that produces the dappled Dachshund. This gene is dominant over nonmerle (m). Where two genes for merle (M) are present, the animal is pure white or nearly so, usually deaf, and often blind as well. Since one-fourth of the progeny of all matings of merle $\times$ merle $(Mm \times Mm)$ should be of this defective type, the ordinary procedure is to cross merle with nonmerle.

The black background of Collies with tan-points $(a^t a^t)$ is the condition in which the merle pattern is most clearly expressed. It consists of alternating and contrasting areas of light blue-gray and of black in different positions, shapes, and sizes.

In sables which are also merled $(a^y a^y Mm$ or $a^y a^t Mm)$ it is often very difficult to see the merle pattern, even though it is potentially and genetically the same as in tricolored individuals. The difficulty in discerning the merle pattern seems to be optical: there is little visible contrast between the lighter- and darker-tan areas of the sable dog.

The basic varieties recognized are as follows:

$a^y a^y BBmm$ —sable, usually clear,

$a^y a^y BBMm$ —sable, usually clear but genetically merle,

$a^y a^t BBmm$ —sable, usually dark and/or shaded,

$a^y a^t BBMm$ —sable, usually dark and/or shaded, apt to show merle pattern especially at birth,

$a^t a^t BBmm$ —black, with tan-points,

$a^t a^t BBMm$ —blue merle, with tan-points,

$a^y a^y BBMM$—white or nearly white, often defective (deaf and/or blind, etc.),

$a^y a^t BBMM$—similar to $a^y a^y BBMM$,

$a^t a^t BBMM$—similar to $a^y a^y BBMM$ and $a^y a^t BBMM$.

Collies have varying amounts of white spotting. The usual condition involves white feet and legs, white chest and brisket, occasionally a white forehead streak, and often a white collar of varying width, sometimes complete and symmetrical, sometimes incomplete. It is not entirely certain whether the white-spotting pattern is due to a well-developed and unusually extended version of the Irish pattern (s^i) or whether it is due to the gene for piebald spotting (s^p). The relatively symmetrical and accurate form of the usual white areas favors the former interpretation as s^i is much more regular in its effects than is s^p, which tends to splash irregular spots in a hit-or-miss fashion.

The gene for extreme-white piebald (s^w) also seems to be relatively common in Collies. In the $s^w s^w$ type a pure-white animal, or one with more or less pigment on the head results. Selective breeding can probably "fix" a type with symmetrical, colored head spots. This has been accomplished in some other breeds. These white or nearly white specimens are admitted by the Standard. They are usually normal in hearing and in vision, thus differing from the defective whites produced by the *MM* combination of genes. The distribution of white on Collies is further discussed under the section on Irish spotting in Part Two of this book.

Occasional blue animals due to the action of the dilution factor (dd) have been reported by breeders. True dilution is rare in this breed but, once encountered, should be obtainable in any of the basic color types that are *DD* or intensely pigmented. In the whole sable series the *dd* forms would differ from the *DD* by being duller, flatter, and perhaps a little lighter in color. It might be difficult to distinguish between the intense and dilute types in this color series. In the tricolors and blue merles ($a^t a^t$) the *dd* animals would be more easily recognized, being anywhere from Maltese blue to pale gray in coloring.

The genes present in this breed are as follows:

DOBERMAN PINSCHERS

All Dobermans thus far recorded have the tan-point pattern ($a^t a^t$). When they have black pigment in full intensity, they are black with tan-points and $BBa^t a^t DD$ in genetic formula. Variation from this black, tan-pointed type occurs in two directions. The first is by the change of black B to brown (liver) b. The brown, tan-pointed variety is $bba^t a^t DD$ in constitution. A change in D, the gene for intensity of pigmentation, may also occur to the dilute (blue-with-tan-points) form. The animals of this type are $BBa^t a^t dd$. These three color varieties are allowed by the Standard, but a fourth variety, which is the natural by-product of crosses between the others, is not accepted.

If a blue $BBdd$ is crossed with a brown $bbDD$, blacks are recreated, $BbDd$ in constitution. If two of these blacks or any blacks like them in genetic formula are crossed together, four color varieties may be produced:

BD—black, with tan-points,
bD —brown (liver), with tan-points,
Bd —blue, dilute black, with tan-points,
bd —dilute brown, with tan-points.

The last-named variety is not mentioned in the Standard. It is rarely seen and probably varies in depth from dark shades to very pale types. If one wishes to avoid confusion *genetically*, this color variety should not be called *fawn*. It would be better to use the terms employed by breeders of Weimaraners, which breed it resembles genetically except for the fact that the Dobermans have a^t (tan points) instead of being A^s (uniform-colored). It is also possible that c^{ch}, a gene producing pale coat color, may influence the depth of pigment in both this and the liver-and-tan variety.

The genes present in this breed are as follows:

$$a^t \quad B \quad C \quad D \quad E \quad g \quad m \quad S \quad t$$
$$ \quad b \quad c^{ch} \quad d$$

GERMAN SHEPHERD DOGS

Accurate predictions of coat color in this breed are made difficult by several circumstances. First, there is great variation in coat color among the "sables," grays, and black-and-tans. The sables, which are $a^y a^y$ or $a^y a^t$, may be almost clear tan, yellow, or cream with a few dark hairs scattered through the coat, or an intermediate grade of color with more black. These may overlap with the lightest extreme of the black-with-tan-points ($a^t a^t$) type. Second, the probable presence of a^w (wild gray-wolflike) coat color, with or without a^y or a^t influences, confuses the situation still further and produces more intergrades that are not distinguishable as genetic types by gross observation.

The segregation between solid blacks and the various types of sables, grays, and blacks with tan points is usually clear and complete. Some blacks may show a few light hairs scattered in the coat, but genetically all blacks are typically A^s (solid color, dominant over all other colors).

There is enough variation between individuals in the depth and richness of the tan (red) pigment to encourage the belief that both the C and c^{ch} genes are present in the breed. The completely white animals recorded may be $a^y a^y c^{ch} c^{ch}$ or, if the ee type of red pattern occurs in the breed, they may be $ee c^{ch} c^{ch}$.

The mask pattern may be present in some animals, independently of whatever allele in the A series the individual possesses.

Truly spotted individuals, i.e., Irish or piebald, do not ordinarily occur. Any white spots that appear are usually small, and these are undoubtedly due to modifiers of the gene S for solid coat color. Mr. Clarence Pfaffenberger of Guide Dogs for the Blind reports a puppy with clear Irish spotting (s^i) very like that seen in Collies. This puppy occurred in pedigreed and registered stock from the mating of two typical nonspotted parents. It represents the result of a mutation carried by both sire and dam.

German Shepherds have the following genes:

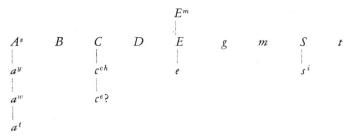

Unpublished data (Pfaffenberger and Ginsburg), soon to be included in a new book by those authors, have given interesting evidence of a recessive form of black coat color in this breed. Actually these black animals usually have a few black hairs between the toes.

They appear to be due to a new allele in the A series which for the present we may designate as a^b. This allele stands at the bottom of the A series and can therefore be carried by parent animals of any of the other colors in the series including tan points (black and tan).

There is also evidence that the presence of a^b in animals heterozygous for it influences the development of the "mask" gene E^m by causing a relatively greater degree of this pattern's extension from the muzzle on to the head. The details of modification of the "mask" pattern development, when they are published, will be of great interest to all breeders.

The same two investigators have also obtained evidence that white coat color in this breed is due to the S^w allele which can be carried as a recessive by certain self or "Irish" spotted individuals.

GREAT DANES

Postponing for the moment consideration and discussion of the harlequin and blue (dilute) colors in Great Danes, let us examine the genetic behavior of black, brindle, and fawn. For some years these have been considered as alternative to one another. They have been classed as three extension grades of black pigment, E, e^{br}, and e. Blacks are representative of full extension or E. Brindles (e^{br}) have partial extension of black, as represented by dark stripes on a fawn background. Recently evidence has been obtained suggesting that the fawns may be due to the a^y mutation rather than to an allele of E. As yet no experimental evidence known to the writer definitely establishes the correct interpretation. In order to establish this a cross should be made between a fawn Dane and a red or yellow of any depth of color from either the Pointer or Setter breeds. If the pups are all red, fawn, or yellow, the ee hypothesis is correct. If they are black, sable, or any dark color, the other hypothesis ($a^y a^y$) is necessary. In this type of cross, made with a fawn Boxer, black is recreated, thus placing the Boxer fawn in the a^y group.

A cross between a fawn Dane and a dark-red Saint Bernard has given only shades of fawn (red), indicating that the a^y gene is present in both breeds.

Some animals have black masks. This marking occurs in either fawns or brindles. The probable genetic nature of the mask pattern has been discussed in Part Two under the E locus.

The dilute series seen most commonly as blues is formed from the intense or fully pigmented varieties by the change of the gene D to d. Thus, intense blacks are $A^sA^sBBDDEE$ in formula, while blues are $A^sA^sBBddEE$. In the brindle series dilute animals are not favored; they are, therefore, rarely seen as adults. When they do occur, they are washed-out or "slaty" in appearance. Dilute fawns also have that flat tint; they are sometimes seen in the show ring. Since the black mask on the muzzle becomes blue, the attractive contrast between the color of that area and the rest of the coat (which is seen in intensely pigmented forms) is lacking.

Harlequins are the most complicated of the color varieties to describe. All have M, the dominant gene for merle. As in other breeds, this produces a distribution of black and light-blue-gray patches on a white background. It is easy to confuse this white background in harlequins with the white spotting caused only by piebald (s^p) and extreme-white piebald (s^w). Actually the pattern in harlequins is probably due to the combined action of M and one of the genes in the s^p, s^w series. It is probable that most harlequins and the merles which accompany them as sibs are $A^sA^sMms^ps^p$, although those with most of the coat pigmented may be $A^sA^sMmSs^p$ in constitution. Other merles without white or with very little may be A^sA^sMmSS.

When the merle pattern is absent and the dogs are mms^ps^p, a typical black-and-white piebald is produced. The shape and distribution of the pigmented spots are, however, usually too extensive and too regular in outline to be considered good harlequins. The "torn" or ragged shape of pigment spots desired in harlequins is a by-product of the action of M, and since this gene also produces gray spots, the breeder of harlequins who tries to meet the Standard is really forced to work against himself by seeking the simultaneous presence of two mutually incompatible characteristics.

Genes found in Great Danes are as follows:

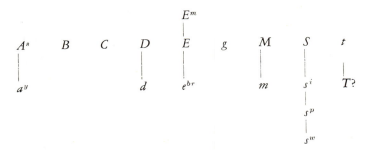

MASTIFFS

The basic colors in this breed are brindle and fawn, the latter being the more common. Black masks are required, but they are absent in some animals.

The data, insofar as they are available, support the conclusion that coat-color inheritance in Mastiffs is similar to that observed in Boxers with the exception of the silver-fawn variety described below. This applies to the basic colors. There are no instances, so far as our records are concerned, of the occurrence of white or nearly white Mastiffs, but otherwise the discussion under Boxers will apply to Mastiffs.

In addition to the brindle and apricot-fawn series, silver-fawn is listed in the Standard for Mastiffs. These varieties are also described in Pugs. The apricot-fawn is probably the ordinary reddish fawn. The silver-fawns may be due to the chinchilla gene (c^{ch}), or even to some more extensive form of pigment change in the direction of albinism occurring in the C series of alleles.

The genes which occur in this breed are as follows:

NEWFOUNDLANDS

The basic color of this breed is deep, solid black. The Standard, however, allows a slight "bronze" tint in the black. This is probably due to fading or some other nutritional or environmental factors rather than to any genetic cause. Animals with white on the chest and/or toes frequently occur. These undoubtedly represent modifiers of the solid-color (self) gene (S) or they may be Ss^p animals.

Landseers usually are genetically piebald because of the s^p gene. The only genetic difference between black Landseers and solid-black Newfoundlands is the presence of s^p in Landseers.

Landseers may also be either solid-colored (other than black), piebald red (bronze), gray (probably blue d), and even pure white (probably s^w; extreme-piebald).

Whitney has reported both blues and reds in Newfoundlands, and the Laboratory's records from cooperators contain one case of a "silver," probably a blue (dd), animal. These should be classed as Landseers.

The nature of the red is uncertain. It may be due to ee, the elimination of black from the coat, or to the a^y gene. Outcrosses with breeds of a known genetic type of red are needed to settle the question.

Whitney (1947) and Leighton (1908) report the occurrence of ticking (T) in white areas of Newfoundlands and state that this is transferable to the white areas of Landseers in crosses.

A good example of the tan-point pattern was observed in a specimen at a large eastern bench show. The tan areas were straw-colored and extensive on the legs, but very little light pigment occurred on the face.

The genes carried by Newfoundlands (Landseers) are as follows:

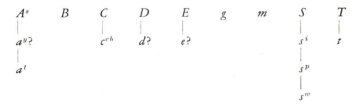

OLD ENGLISH SHEEP DOGS

As in most breeds of long-haired dogs, analysis of the genetics of coat color in Old English Sheep Dogs is not easy, but some facts are clear enough to give us useful information.

Discrimination against brown and fawn of any shade has resulted in establishing the genes B (black), E (extension of dark pigment), and A^s (solid-colored coat). The merle gene (M) is probably present in some individuals, and this complicates attempts to analyze the origin of white spotting, which is frequently encountered. Since solid-colored nonspotted animals are mentioned in the Standard, the gene *S* undoubtedly exists. Dogs with the Irish pattern are numerous. Whether they are the result of the Irish gene (s^i) or are *S* animals carrying piebald (s^p) is difficult to determine.

It may be possible that some sheep dogs are dilute (*dd*) in color, but the fact that puppies are born black or nearly so indicates that dominant dilution (*G*), like that seen in Kerry Blue Terriers and in Poodles, is the usual type. Evidently rigid selection for this lighter coat color has produced animals close to uniformity within the breed.

The genes in this breed are as follows:

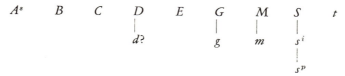

ROTTWEILERS

This breed, calling for a single color, is relatively simple genetically. It has the genes for *B* (black), *E* (extension of dark pigment), *C* (full depth of pigment), *D* (intensity of pigment), and *S* (complete pigment coverage of the coat). It also has $a^t a^t$, the gene for the tan-point pattern. When white mismarking appears, it is probably due to modifying factors acting on the *S*-self pattern.

The genes observed in this breed are:

$$a^t \quad B \quad C \quad D \quad E \quad g \quad m \quad S \quad t$$

SAINT BERNARDS

Although the Standard describes various shades of red as well as "light or dark barred brindle," it seems likely that the whole color range consists of variations in the sable (a^y) pattern, with the added possibility that the black-masked variation of the *E* series of genes is present to complicate matters further. The "brindle" is probably a type of sable (a^y) and is not genetically the same as the true brindle seen in Boxers (e^{br}).

The Standard's description of the "absolutely necessary" white areas coincide with those typical of the Irish (s^i) pattern. (White muzzle, blaze, chest, legs, and tip of tail are specified.) A black mask is desirable but not required. White spots of distinctly greater extent than those specified (characteristic of the piebald (s^p) gene) are commonly seen. Ticking due to gene T is present in some individuals.

The genes in this breed are as follows:

SAMOYEDS

The Standard, which allows "pure white, white and biscuit, and cream," gives the clue to the genetic situation in this breed. Two genes seem to be chiefly involved. The first is extreme-white piebald ($s^w s^w$), which confines pigment, when present, to the ears or to a small patch at the base of the tail. The other gene is probably $c^{ch}c^{ch}$ or $c^e c^e$, which acts particularly on red or yellow pigment, reducing it to "biscuit" or cream. The extreme degree of action by s^w and c^{ch} combined would produce pure white.

The genes in Samoyeds are as follows:

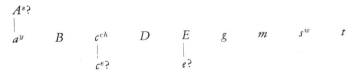

SCHNAUZERS (GIANT)

The solid-black and black-and-tan varieties are Standard examples of known genetic types. The former has A^s for self coat and E for extension of dark pigment. The latter is a^t and E.

The "pepper and salt" variety cannot be definitely classified until outcrosses have been made. It is either brindle, which is a modification of the extension pattern (e^{br}), or wild color, which is a^w, a modification in the A series.

Crosses with a breed such as Boxers, which are fawn a^yE, should help to decide this problem.

The genes in Giant Schnauzers are:

$$A^s \qquad B \qquad c^{ch} \qquad D \qquad E \qquad g \qquad m \qquad S \qquad t$$

$$a^w?$$

$$e^{br}?$$

$$a^t$$

SHETLAND SHEEP DOGS

In this breed the main color varieties and their genetic behavior are apparently identical with those of Collies. Because the genetics of Collies has been discussed in some detail, the reader is referred to that section for information bearing directly on "Shelties."

The following facts may also be of interest. Among the Shelties raised at the Jackson Laboratory several tricolors have been Maltese blue and pale tan instead of being black and tan. They have occurred in sufficient numbers and with enough predictable regularity in certain matings to suggest that they are a true blue dilution (dd) instead of the normal intense pigmentation (DD) which ordinarily characterizes the breed. Co-operators' blanks indicate that a similar condition exists with Collies.

It is also worthy of note that the type of white or nearly white animal that one finds in Collies as an extreme of the piebald series (s^w) has not been so popular in "Shelties," although it occurs in the latter breed.

Evidence of a mutation from a^y or a^t to A^s has been obtained from one mating of Shelties at Bar Harbor. The fact that more than one A^s pup has been produced from one mating, however, makes further study desirable before the mutation hypothesis is accepted.

The genes in Shetland Sheep Dogs are as follows:

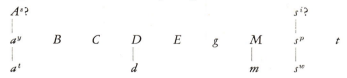

WELSH CORGIS—CARDIGAN

Reds, classified as "sable, fawn and golden" (a^y) and brindle (e^{br}), are the common basic colors. Black with tan-points (a^t) is also found. Blue merles ($a^t a^t Mm$) are listed, as are black and white. The latter may be A^s, representing true genetic blacks, or they may be a^t, tan-point patterns with white occurring in the tan areas. They are not common.

The preference for white markings has resulted in the perpetuation not only of dogs with the Irish pattern but also of dogs with more white, a wide collar or side patches, which the piebald gene (s^p) produces. Occasionally self-colored (S) animals are seen.

The genes in this breed are as follows:

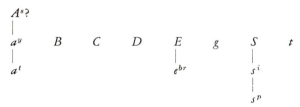

WELSH CORGIS—PEMBROKE

In this breed, the ordinary colors are various shades of sable (red) and black with tan-points. Merles or brindles are rare or absent. The degree of white spotting is almost always restricted to the Irish pattern—white legs and/or paws, chest, and nose spot or blaze.

The genes in Pembroke Corgis are as follows:

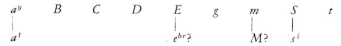

Terriers

AIREDALE TERRIERS

BORN black with tan-points that become lighter with increased tan areas, animals of this breed are undoubtedly $a^t a^t$ in genetic constitution. The individual is usually free from white spots, being solid-colored SS or having only small white patches, which may occur as the result of modifying factors acting on the gene S. In at least one family a mutation directly to piebald spotting (s^p) is known to have occurred. This case has been reported by the writer (1920) and is of great interest because of direct observational evidence and certainty that no mismating could possibly have occurred.

Whitney (1947) believes that the Airedale type of coloration differs from the black and tan of such breeds as Dachshunds, in which there is no progressive increase in red areas and shrinkage of black areas with age. He supposes that a "saddle" factor or gene is present in the former and is lacking in the latter. In view of the very considerable degree of variation in the relative amounts of black and tan within the Airedale breed itself, it would seem that modifying genes acting on the a^t pattern rather than a single gene for saddle may be the explanation. Until further data on crosses between Airedales and other black-and-tan breeds are available, it seems wise to consider the case as being of a doubtful nature.

Airedale Terriers have the following genes:

$$a^t \quad B \quad C \quad D \quad E \quad g \quad m \quad S \quad t$$
$$s^p$$

BEDLINGTON TERRIERS

The pale or dilute degree of pigmentation characteristic of all the color varieties of this breed is not due to the same genetic dilution (d) that one sees in Great Danes, Greyhounds, or Chow Chows. Blue Bedlingtons are born black, and the blues with tan-points are born black with tan-points. These become lighter in color as they develop. The same is true of the liver pups and of the sandy pups which are born liver. Color changes have usually been completed by 6 months of age, and topknots are white by this age. As far as we know, the cause is the G factor for dilution, and it is dominant over its absence. It is found in Poodles, Kerry Blue Terriers, at least some Old English Sheep Dogs, and probably a number of other breeds.

The decrease of pigment in Bedlingtons is perhaps most marked in the region of the topknot, which is very pale or even white. There are solid-colored blue, liver, and sandy, and there are patterns in each of these colors that show lighter pigmentation where the tan of the tan-point pattern is located. We may therefore conclude that both the A^s gene for solid coat and the gene a^t for tan points are found in the breed.

The fact that on maturity some of the sandy animals may be cream-colored and some blues very light suggests that a further reduction due to the gene c^{ch} may occur. It is probable that rigid selection for the paler colors has made the breed essentially uniform for the gene G.

There are few if any records of clear-white spots in adults, so that the breed is probably uniformly SS. If the small white spots seen at birth are retained, it may be that s^i (Irish spotting) is also present.

The ordinary color varieties would then have the following genetic formulas:

Blue—$A^sA^sBBEEGG$; could be A^sA^t or Bb or Ee with no visible difference in color.

Liver—$A^sA^sbbEEGG$; could be A^sa^t or Ee with no visible difference in color.

Sandy—$A^sA^sbbEEGGC^{ch}C^{ch}$; could be $A\,a$ with no difference in color. May be almost cream color.

Blue and tan—$a^ta^tBBEEGG$; could be Bb or Ee with no difference in color.

Liver and tan—$a^ta^tbbEEGG$; could be Ee with no visible difference in color.

Sandy and tan—$a^ta^tbbEEGGC^{ch}C^{ch}$; may be cream.

The genes in this breed are as follows:

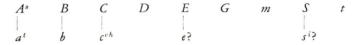

$$\begin{array}{ccccccccc} A^s & B & C & D & E & G & m & S & t \\ | & | & | & & | & & & | & \\ a^t & b & c^{ch} & & e? & & & s^i? & \end{array}$$

BULL TERRIERS

There are two main varieties, white and colored. These are treated as distinct breeds for all practical purposes. The chief and perhaps the only significant difference as far as basic coat-color genes are concerned is that the white variety is s^ws^w or extreme-white piebald. Such color as may occur is therefore ordinarily confined to small eye or ear patches. These are often so restricted in area that the classification of the animal is difficult or impossible, but it is known that these spots may be of any one of three colors, black, brindle, or tan.

In neither the white nor the colored variety are brown or light noses encouraged. It is fair to assume, therefore, that the liver or brown mutation is lacking in the breed and that all animals are BB in genetic constitution.

Colored Bull Terriers are solid-coated (SS or Ss^i) or else they have Irish spotting (s^is^i), with the white confined to the feet, legs, chest, head, or collar.

Briggs and Kaliss (1942) in an analysis of coat-color inheritance in this breed list black, brindle, black and tan, and a series of reds and fawns. The genetic nature of the latter is unknown; they may be of the a^ya^y or ee types, with the chances favoring the former. They also note that blue (probably dd) animals were recorded forty years or more ago. The reader is advised to refer to Briggs and Kaliss' paper, which is helpful to breeders.

Bull Terriers have the following genes:

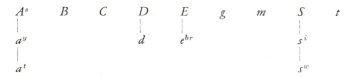

CAIRN TERRIERS

In this breed there are two complications which tend to make definite classification of the coat color very difficult. The first are the color changes that often occur, either for a period up to maturity or even throughout the lifetime of the individual. The changes are usually, but not always, in the direction of increasing lightness of pigmentation. This is very characteristic of the a^y gene in other breeds. The second is the hair length. Patterns such as brindle, tan-points, or sable are much more confused and indistinct on the long-haired breeds. Cairns are no exception to this rule.

The great majority of Cairns are some type of brindle (e^{br}). The darker pigment is uniformly black, but the ground color of lighter pigment varies through a whole series from deep rich red, tan, or dark gray to pale gray-cream or almost white. Animals of the last two colors are probably c^{ch} (pale pigmentation) in character.

Cream and wheaten dogs present a problem in genetics not yet experimentally solved but possible of explanation on the following hypothesis: If the e^{br} gene mutated to E and two brindles which were $a^ya^yEe^{br}$ were mated, there would be a^ya^yEE pups among the progeny. These in all probability would be wheaten. If in addition the c^{ch} gene reduced the tan pigment markedly, creams or even whites might be the result.

There is a strong possibility that the G factor for dilution is also present, but as yet its existence in the breed is not clearly established. White spots, if present, are usually very small and probably represent modifying genes acting on the solid-coat SS type. Black masks are relatively common in brindles and creams and are favored. The black-and-tan (a^t) variety has also been reported. Since this variety militates against dark points, it is not preferred and is rare.

This breed has the following genes:

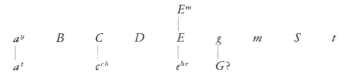

$$
\begin{array}{ccccccccc}
 & & & & E^m & & & & \\
 & & & & | & & & & \\
a^y & B & C & D & E & g & m & S & t \\
| & & | & & | & | & & & \\
a^t & & c^{ch} & & e^{br} & G? & & &
\end{array}
$$

DANDIE DINMONT TERRIERS

The two color varieties of this breed are described as "pepper" and "mustard" when the animals are adult. Both are born darker in shade than their final coat color. The peppers at birth are black with tan-points, while the mustards are dark sable. The fact that the adults are distinctly paler indicates that the dominant dilution factor G is present. There is clear evidence that many, if not all, pepper animals in the breed are genetically $a^t a^t$. Such animals have tan-points. Even the mustards may show traces of this pattern, at least temporarily (Shaw, 1881).

Leighton's *The New Book of the Dog* (1908) mentions the fact that mustard pups have a good deal of black at birth and that, while two mustards bred together frequently produce peppers, two peppers rarely if ever produce mustards. He also notes that crosses between mustards and peppers make the former color darker. These observations very clearly indicate that mustards are $a^y a^y$ when light and $a^y a^t$ when dark. It may also be possible that a change to A^s has occurred in certain animals. This would produce peppers with no trace of tan-points.

The very light color of the topknot is an interesting and as yet unexplained genetic problem. A similar condition is seen in Bedlingtons. Crosses with solid-colored, short-haired breeds carried to at least the second hybrid generation might help elucidate the situation.

A white chest spot is often found in Dandies and probably represents the effect of modifying genes acting on the solid coat SS.

In Dandie Dinmonts the genes are as follows:

$$A^s?$$
$$|$$
$$a^y \quad\quad B \quad\quad C \quad\quad D \quad\quad E \quad\quad G \quad\quad m \quad\quad S \quad\quad t$$
$$|$$
$$a^t$$

FOX TERRIERS (SMOOTH AND WIRE–HAIRED)

The usual pattern in this breed is tricolor $a^t a^t$, although there are a number of animals recorded either as black and white or tan and white. The black-and-white pattern can theoretically occur in one of three ways. It can represent true mutations from a^t to A^s and thus lose the tan-point pattern. It can be tricolor in which all the tan areas chance to coincide with white spots, which thus make the tan invisible. They could also be formed by such an extreme reduction in the area and the depth of the tan that it became impossible or difficult to distinguish tan from black.

Tan-and-white animals can be formed by the mutation of a^t to a^y or by the mutation of E to e. The former would be apt to be shaded sable and the latter clear tan in color. Tans can also result if all the ordinarily black areas happen to be unpigmented because they fall within the white portion of the spotted pattern.

Since the Standard favors a predominance of white, the usual pattern is piebald s^p or extreme-white piebald s^w. The two types of spotting probably overlap in certain individuals, thereby making genetic identification difficult.

There is one well-documented case of a mutation from B to b, producing, in a given line, recurring instances of liver-and-tan tricolors instead of black-and-tan tricolors.

Ticking due to the gene T is commonly found in the smooth-haired variety and may be present as pigmented skin spots with an occasional area of pigmented hair in some Wire-haired Terriers.

The following genes are present in this breed:

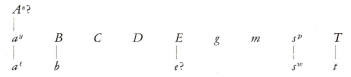

IRISH TERRIERS

These terriers are a sable-tan (red) $a^y a^y$ breed in which the pups are born a darker shade than the clear red which they later become. Variations in the shade of red occur and may be due to c^{ch}, although it is more likely that they are due merely to modifying genes. White spots are due to modifiers acting on the gene S for solid coat color.

Irish Terriers have the following genes:

$$a^y \quad B \quad C \quad D \quad E \quad g \quad m \quad S \quad t$$
$$\qquad\quad \underset{c^{ch}?}{|}$$

KERRY BLUE TERRIERS

Pups of this breed are born black, $A^s A^s$. Some remain black as adults, but others become blue gray of any shade from a very deep slate color to light silvery blue. This is the characteristic behavior of the dominant paling or diluting gene G, which is also found in Poodles, Old English Sheep Dogs, and probably several other breeds. It may be that the more desirable shades of blue are found in Gg animals. If this is the case, black gg individuals will continue to appear, and they can be crossed with light blues to get intermediate shades of blue coat color by incomplete dominance of G.

The rarity of white spots and their small size when they are found indicate that the breed is SS and that modifiers cause whatever spotting there may be.

Breeders have recorded the incidence of the tan-point pattern (a^ta^t), which is not favored by the Standard. Tinges of reddish brown, however, are common. Such tinges do not necessarily mean that an animal is a^t, and they usually disappear with increasing age.

Wheaten or cream individuals, which are probably a^ya^y, have been reported, but they are discarded.

This breed has the following genes:

$$A^s \quad B \quad C \quad D \quad E \quad G \quad m \quad S \quad t$$
$$| \qquad\qquad\qquad\qquad\qquad\quad |$$
$$a^y? \qquad\qquad\qquad\qquad\qquad g$$
$$|$$
$$a^t$$

MANCHESTER TERRIERS

Genetically this is a relatively simple breed as far as coat color is concerned. All animals allowed by the Standard are black with tan-points, a^ta^tBBSS. White spots are undesirable and represent minor departures from complete coverage of the body by the action of the gene S.

Genes of this breed are as follows:

$$a^t \quad B \quad C \quad D \quad E \quad g \quad m \quad S \quad t$$

NORWICH TERRIERS

In this breed the usual color is clear red a^ya^y, occasionally showing persistence of a small amount of dark pigment. In the latter case, the animals are sometimes described as grizzle.

Blacks with tan-points (a^ta^t) are a rarer type and probably have occurred by mutation or by crossbreeding in the origin of the breed. They are recessive to red, and a mating of two black-and-tans should produce only black-and-tan puppies.

All white spots, as in many other terrier breeds, are due to the action of modifying genes on S, the gene for solid coat color.

The following genes are found in this breed:

$$a^y \quad B \quad C \quad D \quad E \quad g \quad m \quad S \quad t$$
$$|$$
$$a^t$$

SCHNAUZERS (MINIATURE AND STANDARD)

Solid blacks (A^s) are found in these breeds. They are probably dominant to the more usual salt-and-pepper variety, although more data on actual litters are needed.

Salt-and-pepper animals are usually nearly black with cream points at birth; they become lighter with banded hairs as they develop. The fact that they are evenly banded suggests that they may have the wild-coat-color pattern (a^w) seen also in Norwegian Elkhounds, where the longer hair gives it a different appearance in adults.

Blacks with tan-points ($a^t a^t$) have been recorded in the Miniatures, but they are not common. The tan in these individuals is usually very pale. This fact, coupled with the almost-white areas in the banded hairs and the cream color of light-pigmented areas in newborn pups, indicates that either c^{ch} or c^e has replaced C, the gene for full, deep pigmentation, in these two breeds.

White spotting, when present, is due to modifiers acting on gene S.

Genes in Schnauzers are as follows:

SCOTTISH TERRIERS

Various shades of brindle, wheaten, or black are the usual colors of this breed. The brindles may be any shade of red or gray as far as the lighter ground color is concerned. The darker shades are probably found in animals that are CC, the intermediate shades in the Cc^{ch} type, and the very pale shades in

$c^{ch}c^{ch}$ individuals. The same relative shades of wheaten are probably to be explained genetically in the same way.

Theoretically the blacks may be of either of two types. They probably are extremely dark or reduced brindles in which no easily distinguishable evidence of the pattern exists. Such blacks would breed like brindles. Black could also occur as the result of mutation from a^y, the gene which is characteristic of brindles, to A^s. Blacks of this sort could carry brindle, being $A^s a^y e^{br} e^{br}$, or they could be $A^s A^s e^{br} e^{br}$ and breed true.

In the latter part of the nineteenth century and the early years of the twentieth the black-with-tan-points variety became less common; it has now largely disappeared from most of the strains being used to produce bench-show animals.

An instance of a mutation from black B to liver b was recorded in an Iowa kennel some forty years ago, and the writer saw two of the liver-brindle (bb) animals with light-yellow eyes and flesh-colored noses. Otherwise they were typical Scottish Terriers in conformation. The writer has also observed a mutation from the ordinary solid-colored type S to piebald s^p in a carefully controlled mating in his father's kennel. This piebald animal has been shown in a paper by the writer (1920).

Light-yellow or cream and pure-white Scotties (formerly known as Roseneath Terriers) represent a wheaten type in which the gene C for full pigmentation has been replaced by the paling gene c^{ch}. The mutation is a rare one but has been observed by the writer in his father's kennels. Older dog books describe the relationship of creams and whites in a manner consistent with the above theory.

Wheatens, like sables in other breeds, may be clear-colored or may show varying amounts of scattered dark pigment, especially on the back and sides. Usually this disappears with age.

Masks may well have been present at one time. They are favorably mentioned (Shaw, 1881). It is interesting to note that they are still found in Cairn Terriers, a breed which probably has common ancestors with the Scotty.

In Scottish Terriers the genes are as follows:

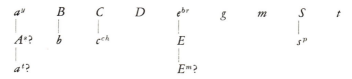

SEALYHAM TERRIERS

This breed seems, by selection, to have come to include only s^w (extreme-white piebald) animals, many of which are clear white. The color spots are so restricted that it is often hard to classify the genetic type involved. It would appear that sable-tan ($a^y a^y$) is the usual type. Under this pattern one would expect "lemon, tan and badger" which the Standard mentions.

This breed has the following genes:

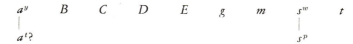

SKYE TERRIERS

Although the longer coat produces a less-distinct type of brindle than when e^{br} is found in shorter-coated breeds, the various shades of that pattern and of wheaten resemble the colors and follow the genetic behavior of Cairn Terriers.

Solid-colored dogs, which may be A^s with no trace of brindle, are probably gray in any shade from dark to light. This is because of the action of the graying gene G on a long-haired breed.

Liver due to b instead of B has been reported but is not common. Some individuals have a light topknot very much like those of the Bedlington or Dandie Dinmont. The sections on those breeds should be read with that fact in mind. It is uncertain whether or not the gene E^m for black mask is present.

Skye Terriers have the following genes:

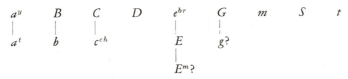

WELSH TERRIERS

This is a black-and-tan $(a^t a^t)$ breed with occasional individuals showing an admixture of lighter hairs in the dark areas producing "grizzles." The extent of the dark pigment varies somewhat, although it is commonly found over most of the body in contradistinction to the much-reduced saddle often seen in Airedales.

In this breed the following genes are found:

$$a^t \quad B \quad C \quad D \quad E \quad g \quad m \quad S \quad t$$

WEST HIGHLAND WHITE TERRIERS

The white in this breed is not due to the action of extreme-white piebald (s^w) as it is in Sealyhams but rather to an advanced degree of paling of pigment (c^{ch}) or to reduction (c^e) in an otherwise solid-colored dog. Rigid selection has fixed the type. This is indicated by the occasional appearance of very pale cream tinges in whole individual hairs or at the base of white-tipped hairs. The probability is that the basic color is very pale wheaten $(a^y a^y)$ modified as above-stated by mutation to one of the extreme diluting genes in the C series.

Most Highland Terriers have the following genes:

$$a^y \quad B \quad \begin{matrix} c^{ch} \\ | \\ c^e? \end{matrix} \quad D \quad E \quad g \quad m \quad S \quad t$$

CHAPTER XIV

Toys

AFFENPINSCHERS

BLACK is the most desired color in the breed and is undoubtedly produced by the A^s gene. "Red, gray and other mixtures" spoken of in the Standard are probably various shades of the tan-sable (a^y) series. Since black with tan-points is also an approved color, we know that the $a^t a^t$ combination exists. This would mean that the darker grays are $a^y a^t$ in formula. The black mask (E^m) is present in some individuals.

The penalty imposed on very light colors means that there has been selection for the C (full pigment), rather than the c^{ch}, gene. Similarly, the objection to white markings means a selection for gene S (solid coat). If any white appears it is due to modifiers of S and not to s^i (Irish spotting).

Genes in this breed are as follows:

$$
\begin{array}{cccccccc}
A^s & B & C & D & E & g & m & S & t \\
| & & & & | & & & & \\
a^y & & & & E^m & & & & \\
| & & & & & & & & \\
a^t & & & & & & & &
\end{array}
$$

CHIHUAHUAS

A great variety of colors occur in this breed. Among the commoner ones are various shades of fawn and sable-tan. Some of these are undoubtedly $a^y a^y$ or $a^y a^t$, for blacks with the tan-point pattern can be produced by two sable or tan parents. Blacks $(A^s E)$ have also been produced by two tan or yellow parents, and this suggests that ee types of reds are also present.

Not only does the whole series of black-nosed types (B) exist, but their liver counterparts in the bb types are also found.

Spotted $s^p s^p$ and Irish $s^i s^i$ spotting are present in addition to the nonspotted pattern. The piebald (s^p) animals vary in amount of pigment from those which resemble animals with the Irish type of spotting at one end of the scale to those which have only one or two small pigmented areas at the other.

Because very pale cream or even white individuals occur, it is probable that the c^{ch} type of pigment reduction is also found in this breed. The gene d for blue dilution has been recorded but is not frequently seen.

One can readily realize that the different combinations of these various color and pattern genes produce such a varied group of color types that the Standard does not attempt to define or describe them.

The genes in Chihuahuas are as follows:

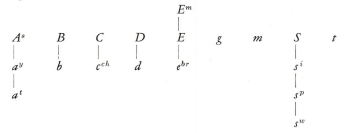

ENGLISH TOY SPANIELS

There are four recognized varieties of this breed, as follows:

(1) King Charles, a bicolor, black with tan-points $(a^t a^t EE)$. This variety must have little or no white and therefore is undoubtedly SS in formula.

(2) Prince Charles, a tricolor exactly the same as the above in regard to tan-points but with piebald $(s^p s^p)$ replacing the gene for solid coat (S) and with E still present.

(3) Ruby, a solid, rich, clear red. Undoubtedly this differs from King Charles merely in the substitution of ee for EE. It has the formula $a^t a^t ee SS$.

(4) Blenheim, a red-and-white piebald which differs from the Ruby merely in the substitution of $s^p s^p$ for SS. The depth of color in all specimens suggests that all have the gene C for full pigmentation.

The genes in these varieties are:

$$a^t \quad B \quad C \quad D \quad \underset{e}{\overset{E}{\mid}} \quad g \quad m \quad \underset{s^p}{\overset{S}{\mid}} \quad t$$

GRIFFONS (BRUSSELS)

The commonest color in this breed is "reddish brown" ($a^y a^y$), with or without a black mask. The black mask is rarely distinguishable in the wire-haired variety but is much more noticeable in the short-coated Brabançon. Blacks with tan-points are occasionally found. These are $a^t a^t$, and when two are crossed together the color pattern breeds true. In Belgian Griffons solid black ($A^s A^s$) is found in addition to the a^y and a^t patterns. The author has published a short note (1934) on the color genetics of a^y and a^t in Brussels Griffons.

Griffons have the following genes:

$$\underset{a^t}{\overset{a^y}{\mid}} \quad B \quad C \quad D \quad \underset{E}{\overset{E^m}{\mid}} \quad g \quad m \quad S \quad t$$

ITALIAN GREYHOUNDS

The Standard allows "all shades of fawn, red, mouse, blue, cream and white." The listing of blue and mouse indicates clearly that blue dilution (d) occurs in this breed. The mention of all shades of fawn and red suggests that the two genetic types of red ($a^y a^y$ and ee) are present. The cream and white varieties are probably the result of the c^{ch} type of paling; the other types are undoubtedly fully pigmented C. No mention is made of black, which theoretically should occur if blue and deeply pigmented fawn are commonly seen.

It is also interesting to note that no brindles are recorded today. This variety and black were reported in the early history of the breed, but the Standard did not favor them and they have been greatly reduced or eliminated by selection.

Solid-colored animals and piebalds occur in any color. Probably Irish and extreme-white piebalds are also found, but they are not common.

The genes in this breed are:

JAPANESE SPANIELS

All individuals of the breed are piebald $s^p s^p$, with clear-cut demarcation of pigmented and white areas.

There are two types of pigment, black and red (yellow). The latter is sometimes of the ee type since the pigment is clear and shows no dark hairs or streaks as is so often seen in the $a^y a^y$ tans and red. Other red animals, however, are probably a^y, and these may show some dark hairs mingled with the red.

This breed has the following genes:

A^s	B	C	D	E	g	m	s^p	t
a^y		c^{ch}		e			s^w	

MALTESE

Pure white is the accepted color. While the genetic nature of the white is not certain, there is strong evidence that it is the result of extreme-white piebald ($s^w s^w$), as in the English Bull Terrier. An occasional individual with a cream spot or cream ears is more likely according to this theory than if one supposes the white to be only the result of ee yellows with extreme action of the c^{ch} (paling) gene.

Since black noses and eyes are required, it is probable that all individuals are *BB* in genetic type. This would make the breed $A^sA^sBBs^ws^w$ in formula.

The genes in Maltese are as follows:

A^s B c^{ch} D e g m s^w t
$|$
c^e?

PEKINGESE

Long hair and a wide variety of color genes make the identification of genetic types in this breed a very difficult matter. One of the difficulties is the use of the term "brindle" to describe a mixture of black and red hairs. Such animals are probably genetically a type of sable a^ya^y or a^ya^t and *not* true brindles due to action of the e^{br} gene.

The occurrence of blacks as the result of matings of red × sable suggests that the two genetic types of red are present— the a^y series (dark sable or red) and the ee series (clear red). Some animals recorded as black at birth become sable with increasing age.

The probability that there are two genetic types of red is further supported by the types of "whites" or "partial albinos" which have been described. Creams with pigmentation still further reduced by c^{ch} or extreme dilution c^e would become one variety of whites. These are probably ee reds which are bleached out by paling genes. True albinos with pink eyes have been reported. The smoky whites or cornaz albinos reported by Pearson and Usher (1929) may be the a^ya^y reds (sables) similarly paled.[1] A point in favor of this is a report of the production of more heavily pigmented animals from a cross of cornaz × white. A third type of white might be the extreme-white piebald variation. These are usually dark-eyed but could some-

---

[1] There is also the possibility that a "pink-eyed" gene like gene p of rodents may be the cause of this color. It is so uncommon, however, that more firsthand data are needed before it can be definitely classified.

times be blue-eyed. It is known that Pekes have all four of the
S-series types—S (solid), s^i (Irish spotting), s^p (piebald), and
s^w (extreme-piebald). Some $s^w s^w$ animals may have blue eyes
and an entirely white coat.

The presence or absence of black masks on red or sable animals is a further complication. It is probable that the presence
of the mask is dominant to its absence.

In the A series of patterns Pekes may be A^s (solid black),
a^y (sable or red), and a^t (black with tan points). The first- and
last-named varieties are comparatively uncommon.

There is one well-authenticated record of a blue Pekingese,
apparently due to the gene d for blue dilution. This gene is
not sufficiently common in the breed to be of great importance
for the breeder.

As yet we have no record of the liver (b) mutation in Pekes,
which is probably just as well since a bb series in addition to
the black-pigmented BB types would at least double the difficulties of classifying individual animals.

The genes in Pekingese are as follows:

PINSCHERS (MINIATURE)

Certain colors are accepted by the Standard for this breed.
One of these, red, is undoubtedly $a^y a^y$ in genetic formula.
Another is black with tan-points, $a^t a^t$.

A correspondent has mentioned two black puppies "without
markings." Both parents were red. One of the black pups was
bred to a black-and-tan and gave 6 pups, 2 of which were jet

black and 4 black with a reddish cast (not tan-point markings). These exceptional blacks pose a difficult genetic problem, for if a mutation had occurred from a^y to A^s, they still should have carried a^y as a recessive. This, in turn, should have produced some reds when the mating with a black-and-tan was made. This report therefore presents a situation unique in the genetics of coat color in dogs. The only possible explanation seems to be that one of the red parents was a light-colored solid liver rather than a true red. Until further data are available, this case will have to remain unexplained.

Brown-and-tans (liver-and-tans) $a^t a^t bb$ are recorded and accepted, as are solid livers. If the latter occur, so should solid blacks in rare cases. Blues and blue-and-tans due to the action of gene d are also known.

Merles (harlequins), undoubtedly due to gene M, have been described by a breeder in a leading dog journal. They were formerly accepted by the Standard but are now excluded. The appearance of large areas of white in harlequins is a complication because piebalds are not recorded. It may be that the harlequins are homozygous MM merles, for it is known that in Dachshunds such animals may have a large part of the coat white.

In this breed the following genes are reported:

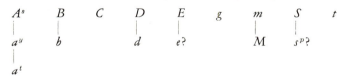

POMERANIANS

In the A-pattern series, the solid-colored (A^s) coat is commonly found. So also is a^y, the gene producing sable or red of various shades. Although the black, tan-pointed pattern (a^t) exists, it has not been frequently reported. It usually produces the typical tan markings on face, chest, feet, and tail, which contrast sharply with the black body color. Occasionally, if the black pigment is reduced in extent to a saddle, the pattern may be almost impossible to distinguish from a very dark sable. The tan-pointed pattern may occur in combination with other colors than black.

The B gene has mutated to b, and this has given rise to a liver (chocolate) series. In the bb types, all varieties of a^y (sables) probably become clear, bright red, and the liver shadings do not stand out clearly as do the black in $a^y a^y BB$ individuals.

Dogs which are $a^y a^y bb$ undoubtedly belong to the orange variety of the breed. As yet there is no evidence of yellow or red coat color due to the ee mutation. Black masks due to E^m are recorded, but they are not generally considered desirable. The depth of red (orange) or of liver (chocolate) pigment varies greatly. Cream animals, some of which are very pale liver (brown or chocolate), are also observed. The lighter shades are usually the result of paling genes, c^{ch} or possibly c^e, extreme paling. The most pronounced degree of paling undoubtedly produces the now very rare white animal.

In addition to the paling due to genes of the C series, there is evidence that d (blue dilution) may exist in some strains. Since this gene would produce maltese blues in BB individuals as well as light liver (chocolate) in bb dogs and since c^{ch} would affect the latter type only and not the B series, there is a chance to estimate the probable causative gene in pale-colored individuals. Blues are much rarer than dilute liver, and this suggests that the d gene is very limited in its distribution throughout the breed.

Piebald (parti-color) patterns are permissible, but they are rarely seen today. The Irish type of spotting is considered mismarking, as it is in Cocker Spaniels, and is discouraged. Since this type of spotting is very apt to appear in solid-colored dogs which carry the recessive piebald (Ss^p) gene, this would mean the gradual decrease in the relative frequency of piebalds ($s^p s^p$) because two Ss^p (Irish-spotted; mismarked) individuals are not bred together as a general rule.

In Pomeranians the following genes occur:

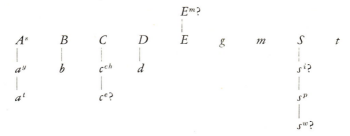

$$
\begin{array}{ccccccccc}
 & & & & E^m? & & & & \\
 & & & & | & & & & \\
A^s & B & C & D & E & g & m & S & t \\
| & | & | & | & & & & | & \\
a^y & b & c^{ch} & d & & & & s^i? & \\
| & & | & & & & & | & \\
a^t & & c^e? & & & & & s^p & \\
 & & & & & & & | & \\
 & & & & & & & s^w? & \\
\end{array}
$$

PUGS

Black pugs are solid-colored (A^s) animals. They are undoubtedly also SS, since white spots are rigidly discouraged.

The apricot-fawns are in the a^y group. They uniformly have the black mask (E^m) and are tan or light red in color, which indicates that they are fully pigmented by the action of the C gene.

The silver-fawns on the other hand show a distinct dullness and paling of the nonblack portion of the coat. This seems to be due to action of the gene c^{ch}.

No liver-colored pugs have been reported, which means that the breed as a whole has the gene B for black pigment.

In Pugs the following genes are present:

$$
\begin{array}{ccccccccc}
A^s & B & C & D & E^m & g & m & S & t \\
| & & | & & & & & & \\
a^y & & c^{ch} & & & & & & \\
\end{array}
$$

TOY MANCHESTER TERRIERS

In this breed only the one color type is allowed. The animal must be black in basic color with deep, rich tan-points. This definitely places it in the $a^t a^t$ group. The insistence on deep tan means that the gene C for full pigmentation is required.

Two other varieties have been reported. One is blue and tan formed by mutation from D to d. The other, red, has been observed in England and by analogy with other breeds is probably ee in genetic formula.

The rigid discrimination against white spots ensures the presence of the gene S.

In this breed the following genes are recorded:

$$a^t \quad B \quad C \quad D \quad E \quad g \quad m \quad S \quad t$$
$$\qquad\qquad\qquad | \qquad | $$
$$\qquad\qquad\qquad d \qquad e$$

YORKSHIRE TERRIERS

Puppies in this breed are born black with somewhat indistinct tan-points. This indicates that the breed is $a^t a^t$ in genetic type. As the animal ages and the hair lengthens, there are two changes in color pattern which occur gradually. First, the area covered by the dark pigment decreases gradually, somewhat as in Airedales, so that it becomes a very heavy saddle rather than a complete body coverage. It is, however, essential that dark pigment cover the nape and extend unbroken to the head. Second, the black pigment pales to a steel-blue, which suggests the action of the dilution factor G seen also in Bedlington and in Dandie Dinmont Terriers.

All Yorkshires carry the gene B for black and S for solid coat, i.e., absence of white spotting.

Yorkshire Terriers have the following genes:

$$a^t \quad B \quad C \quad D \quad E \quad G \quad m \quad S \quad t$$
$$\qquad\qquad | $$
$$\qquad\qquad c^{ch}$$

CHAPTER XV

Nonsporting Dogs

BOSTON TERRIERS

ALL individuals of this breed must have a black nose and no suggestion of liver or chocolate pigment. They therefore have the gene B, not b. They must also be spotted with white, and, whenever possible, the white spots should appear in a regular and symmetrical pattern. Symmetry in spotting is usually a characteristic of the s^i gene for Irish spotting but not of either the s^p gene for piebald or of the s^w gene for extreme-white piebald. It is probable that the pattern of white spotting most desired by the Standard can be formed by the Irish spotting gene $(s^i s^i)$ with minus modifiers or, in certain cases, by a combination $s^i s^p$ or $s^i s^w$.

The breeding records of cooperating fanciers show that white or nearly white individuals do occur. We shall see this again in French and in English Bulldogs. There are also cases of animals with various degrees of irregular piebald spotting. The former type $(s^i s^p)$ would not ordinarily throw irregularly spotted or all-white pups; the latter $(s^i s^w)$ might.

The basic pattern of the dark pigment is brindle (e^{br}). There has been some selection toward darker types, and this has made the light bands indistinct, reduced, or even indistinguishable. The term "seal" is used to describe such a coat color.

Black coats have also been recorded frequently. These might come about by an accumulation of darkening modifiers of brindle, as it does in some Scottish Terriers, or it may be due to a change in the a^y gene, which results in a mutation to A^s. There is no way of distinguishing between the two types of origin except by carefully planned and controlled breeding tests over a number of generations.

In this breed there are the following genes:

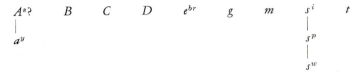

BULLDOGS (ENGLISH)

If the reader will study the description of coat-color inheritance in Boxers, he will obtain most of the necessary information on English Bulldogs. It should, however, be noted that both black (A^s) and black-with-tan-point (a^t) patterns are found in Bulldogs, though not in Boxers.

The breed carries the gene B for black pigment and is normally fully pigmented (C) rather than pale (c^{ch}), although the type of red described as "fallow" by the Standard may well be the c^{ch} degree of pigmentation.

Apparently all four types of the S series, namely, s^w (white or extreme-white piebald), s^p (piebald), s^i (Irish spotting), and S (solid color; no white spots), occur and none of them are penalized by the Standard.

Brindle and red of various shades are the two common basic color types, but solid black has been recorded in the colored patches of piebald dogs. The brindles, reds, or fawns may or may not have black masks. The occurrence of white on or near the muzzle and between the eyes naturally makes classification as to the mask a difficult matter.

Some individuals have such small spots of pigmented hair on a white background that the possibility of ticking (T) must be considered.

The genes in Bulldogs are as follows:

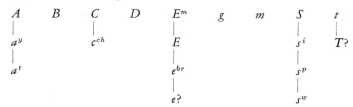

CHOW CHOWS

The presence of long hair makes the accurate determination of the genetic nature of the red, sable, cinnamon, and cream series difficult in this breed. There is, however, a considerable amount of information on coat-color inheritance which can be utilized.

First, most Chows are A^s (solid-colored). Shaded sables (a^y) are allowed but not common. Blacks with tan-points (a^t) are also reported, but these are discouraged by the Standard. Second, both kinds of red (a^y sable and e, failure to form dark coat pigment) are found in the breed. Third, true blue dilution (d) forms the blue and some (if not all) of the cinnamon varieties.

Pale colors may also be the result of the action of the c^{ch} gene. This would explain the very light creams, the cream-tipped whites, the occasional pure whites, and probably the obscure and not yet analyzed variety "silver grey."

All Chows of which records are available are of the B (black-pigmented) rather than of the b (liver) type. All are of the solid-coated (nonspotted) type S except for the very slight and often unnoticed effects of minus modifiers, which may produce a few white hairs or a tiny white spot on the chest, forehead, or feet.

Chows have the following genes:

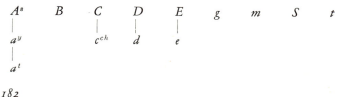

DALMATIANS

The color of the many small pigmented spots in this breed is commonly black over the whole body surface, on the legs, and on the tail. This would indicate A^s for complete-body-coverage pattern, B for the formation of black pigment, and E for its extension. Liver-spotted animals, which are A^sbE, are also fairly common.

Individuals which have black body spots but tan spots on the face and legs have also been observed. These are undoubtedly a^ta^t black with tan-points. Less frequent are the ee yellow (orange or lemon) and white types. These if black-nosed are either A^sBe or a^tBe; if liver-nosed they may be A^sbe or a^tbe.

The characteristic spotted pattern appears to be due to the combined action of two genes, both present in all Dalmatians. The first of these is the s^ws^w type (extreme-white piebald), and this accounts for the newborn Dalmatian being either clear white or marked with perhaps one or two small but distinct patches of pigment. Added to this pattern and expressing itself progressively for some weeks after birth is the dominant ticking gene T which spatters the whole body surface with the characteristic Dalmatian spots.

In Dalmatians there are the following genes:

$$A^s \quad B \quad C \quad D \quad E \quad g \quad s^w \quad T$$
$$a^t \quad b \qquad\qquad e$$

FRENCH BULLDOGS

The coat colors in this breed generally resemble, and are inherited in the same way as, those of English Bulldogs. The varieties of brindles are the same and fawns are similar. According to the Standard, liver, mouse, black with no trace of brindle, and black with tan points constitute a disqualification. A large white chest spot is often seen. This is either a modified Irish pattern (s^i) or minus modifiers on the self-coated (S) variety. The piebald spotting (s^p) is usually irregular, as in English

Bulldogs, and does not show the same degree of symmetry as in Boston Terriers. All-white, or nearly all-white, individuals occur, and these seem to be due to the action of $s^w s^w$, the extreme-white piebald gene.

The ticking gene (T) is present in some individuals.

In French Bulldogs there are the following genes:

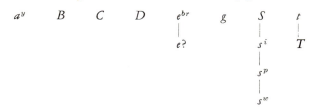

$$a^y \quad B \quad C \quad D \quad e^{br} \quad g \quad S \quad t$$

KEESHONDEN

Pups of this breed are born practically black, with perhaps a few scattered gray hairs. The black mask is usually if not always present. This suggests that the genetic analysis made of the Norwegian Elkhound may apply here as well. On the other hand, the color of Keeshonden resembles at least superficially the so-called wolf-sable in Pomeranians, and it is likely that Pomeranians are $a^y a^y$, with modifiers that increase the amount of dark pigment.

Orange-sable Keeshonden have been reported, which may possibly favor the hypothesis that the ordinary Keeshond is $a^y a^y$ with modifiers and that the orange-sable has lost these modifiers. Further breeding experiments including carefully planned outcrosses are required for complete analysis.

In Keeshonden there are the following genes:

$$a^w \quad B \quad C \quad D \quad E \quad g \quad m \quad S \quad t$$

POODLES

Comments on the inheritance of coat colors in Poodles apply with equal accuracy to all three types, Standard, Miniature, and Toy. They will therefore be grouped together in this section.

The combination of long hair and many coat-color genes results in an extremely complex genetic situation and makes it difficult to distinguish accurately the different types in the blue-gray-silver and apricot-cream series. It will probably be well, therefore, to start with a simple difference, namely, that between the black-haired and black-nosed series having gene B and the liver series where B is replaced by b. In solid-colored, fully pigmented blacks and livers, the two varieties are distinct and have no intermediate type. BB or Bb are black and bb liver (brown).

Black (BB or Bb) animals lose the depth of their pigmentation because of two genes, singly or in combination. One of these genes is the dominant gray-dilution G. $BBGG$ or $BBGg$ pups are black at birth and, like the Kerry Blue Terriers, may gradually become blue or gray as they mature or even throughout life. The exact time of change may vary, but the form of the gene action is comparable in the two breeds. The second gene d is recessive and produces blue dilution, which is present at birth and which does not change with age. Where the individual has both G and d, we should expect the lightest type of animal having any black pigment extended throughout its coat.

Exactly parallel effects of G and d are to be found in the liver (brown, bb) series, with the additional factor that c^{ch} may produce a still paler coat color. It should be remembered that although the gene c^{ch} may be present in, and is transmitted by, blacks, it does not produce any visible effect on that coat color.

We may thus list one series of blacks and two of livers or browns as they progress from dark to light shades.

1.
$BDgE$—intense black	$BdgE$—maltese blue
$BDGE$—grayish blue to silver	$BdGE$—silver

2.
$bDgCE$—intense liver	$bdgCE$—dull bluish brown
$bDGCE$—grayish brown	$bdGCE$—silver, brown cast

3.

bDgc^{ch}E—brown or beige	*bdgc^{ch}E*—very pale smoky beige
bDGc^{ch}E—silver-beige	*bdGc^{ch}E*—"off white" or white

All these forms have the gene *E*, which allows whatever dark pigment there is to appear anywhere on the body surface.

Another complete series of types with *e*, which restricts black or liver to the nose and eyes and which allows only some shade of the red-yellow type of pigment, is also formed. The types included are certain to be confusing and often indistinguishable from one another except by breeding tests. Their appearance is estimated to be as follows:

Black noses	*Brown noses*
BDgeC—light red, *BDgec^{ch}*—cream to white	*bDgeC*—pinkish beige; *bDgec^{ch}*—cream to white
BDGeC—pale, silvery red; *BDGec^{ch}*—cream to white	*bDGeC*—pale silver-beige; *bDGec^{ch}*—white
BdgeC—light dull smoky red; *Bdgec^{ch}*—cream to white	*bdgeC*—dull cream; *bdgec^{ch}*—white
BdGeC—very pale silvery red; *BdGec^{ch}*—white	*bdGeC*—pale cream to white; *bdGec^{ch}*—white

The absence of dark, rich red in Poodles is interesting and needs careful genetic study in order to arrive at an explanation.

Breeders of Poodles speak of two kinds of apricots, those born black and those born apricot. This is a good example of the type of confusion created by using a *single* descriptive term to describe the color of what are clearly two distinct genetic types. Apricots born black almost certainly have the paling gene *G*; apricots born so and staying so would theoretically be without it. Those born apricot and becoming cream *might* well have it. These explanations are frankly estimates, but they have a reasonable basis of observed genetic tests and they should stimulate more extensive and more accurate experimental breeding.

Blacks with tan-points and livers with tan-points are frequently seen. In *a^{t}a^{t}c^{ch}c^{ch}* varieties the tan points may be so reduced in pigmentation as to be very nearly white.

There is clear evidence that the terminology used by Poodle breeders is so confused and complicated that no amount of discussion, correspondence, or citation of colors of litters will bring order out of the chaos that exists. Without doubt the terms "silver," "blue," "taupe," and "gray" are used by one breeder to describe different colors from those similarly described by another. The four color types overlap, produce visually indistinguishable varieties, change with age, and show different degrees of expression according to the recessive colors they carry. Only a long series of carefully planned breeding experiments with definite outcrossing with breeds of known genetic types will solve the problem. Until this is done the genes c^{ch}, G, and d will continue to provide complexities that cannot be solved by debate.

The genes in Poodles are as follows:

A^s	B	C	D	E	G	m	S	t
a^t	b	c^{ch}	d	e	g			

Bibliography

Anker, J. 1925. Die Vererbung der Haarfarbe bei Dachshunds. V. Danske. Vidensk. Selsk. Biol. Meddel, 4, No. 6.

Ash, E. C. 1921(?). Dogs: Their History and Development. Boston, Houghton Mifflin Co.

Barrows, W. M., and Phillips, J. M. 1915. Color in Cocker Spaniels. Jour. Heredity 6: 387–397.

Briggs, L. C., and Kaliss, N. 1942. Coat Color Inheritance in Bull Terriers. Jour. Heredity 33: 223–228.

Burns, M. 1952. The Genetics of the Dog. R. Cunningham & Sons, Alva, Eng. (An excellent bibliography is contained in this book.)

Castle, W. E. 1930. Genetics and Eugenics. Harvard University Press, Cambridge, Mass.

The Complete Dog Book. 1951. Garden City Books, Garden City, New York. (Official publication of the American Kennel Club.)

Darling, F. F., and Gardner, P. 1933. A Note on the Inheritance in the Coloration of Irish Wolfhounds. Jour. Genetics 27: 377–378.

Davis, H. P. 1949. The Modern Dog Encyclopedia. Stackpole and Heck, Harrisburg, Penn.

Dawson, W. M. 1937. Heredity in the Dog. Yearbook of Agriculture, U.S. Dept. of Agriculture, Washington, D. C.

De Bylandt, H. 1905. Dogs of All Nations. 2 vols. Kegan Paul, Trench, Trübner & Co., London.

Doncaster, L. 1905. Proc. Camb. Philos. Soc. 13: 215.

DuBuis, E. M. 1948. A Rose by Any Other Name. New England Dog, July, 1948.

Engelmann, F. 1925. Der Dachshund. Carl Schmitt, Jena.

Gayot, E. 1867. Le Chien. Firman Didot Frères, Paris.

Gray, D. J. T. 1891. The Dogs of Scotland. J. P. Mathew & Co., Dundee, Scotland.

Haldane, J. B. S. 1927. The Comparative Genetics of Colour in Rodents and Carnivora. Biol. Rev. 2: 199–212.

Bibliography

Hubbard, C. L. B. 1952. The Observer's Book of Dogs. Frederick Warne & Co., London.

Humphry, E., and Warner, L. 1934. Working Dogs. Johns Hopkins Press, Baltimore.

"Idstone." 1872. The Dog. Cassell, Petter and Galpin & Co., London.

Leighton, R. 1908(?). The New Book of the Dog. 2 vols. Cassel & Co., London.

Little, C. C. 1914. Coat Color in Pointer Dogs. Jour. Heredity 5:244–248.

Little, C. C. 1920. A Note on the Origin of Piebald Spotting in Dogs. Jour. Heredity 11:12–15.

Little, C. C. 1934. Inheritance in Toy Griffons. Jour. Heredity 25:198–200.

Little, C. C. 1948. Genetics in Cocker Spaniels. Jour. Heredity 39:181–185.

Little, C. C., and Jones, E. E. 1919. The Inheritance of Coat Color in Great Danes. Jour. Heredity 10:309–320.

Marchlewski, T. 1930. Genetic Studies on the Domestic Dog. Akad. Umiefetnosci, Krakow 2:117–145.

Mitchell, A. L. 1935. Dominant Dilution and Other Coat Factors in Collie Dogs. Jour. Heredity 26:425–430.

Morgan, T. H. 1919. The Physical Basis of Heredity. Lippincott, Philadelphia, Penn.

Onstott, K. 1946. The Art of Breeding Better Dogs. Denlinger's, Washington, D. C.

Pearson, K., Nettleship, E., and Usher, C. H. 1911. A Monograph on Albinism in Man. 3 vols. Dulav & Co., London.

Pearson, K., and Usher, C. H. 1929. Albinism in Dogs. Biometrika 21:144–163.

Phillips, J. McI. 1938. Sable Coat Color in Cockers. Jour. Heredity 29:67–69.

Reichenbach, Dr. 1836. Der Hund. F. Ries, Leipzig.

Shaw, Vero. 1881. Book of the Dog. Cassell, Petter and Galpin & Co., London.

Shields, G. O. 1891. The American Book of the Dog. Rand, McNally & Co., New York.

Stockard, C. R. 1941. The Genetic and Endocrine Basis for Differences in Form and Behavior. Wistar Institute of Anatomy, Philadelphia, Penn.

"Stonhenge." 1872. The Dog in Health and Disease. Longmans, Green, Reader & Dyer, London.

Warren, D. C. 1927. Coat Color Inheritance in Greyhounds. Jour. Heredity 18: 513–522.

Whitney, L. F. 1928. The Basis of Breeding. Fowler, New Haven, Conn.

Whitney, L. F. 1928. The Inheritance of a Ticking Factor in Hounds. Jour. Heredity 19: 499–502.

Whitney, L. F. 1947. How to Breed Dogs. Orange Judd Co., New York.

Winge, Ö. 1950. Inheritance in Dogs. Comstock Publishing Co., Ithaca, N. Y.

Wriedt, C. 1925. Letale Faktoren. Z. Tierz. ZüchtBiol. 3: 223–230.

Youatt, W. 1886. The Dog. Longmans, Green & Co., London.

Index